Complexity in a Ditch

Bringing Water to the Idaho Desert

Complexity in a Ditch

Bringing Water to the Idaho Desert

HUGH T. LOVIN

INTRODUCTION BY

ADAM M. SOWARDS

Washington State University Press
Pullman, Washington

Washington State University Press
PO Box 645910
Pullman, Washington 99164-5910
Phone: 800-354-7360
Fax: 509-335-8568
Email: wsupress@wsu.edu
Website: wsupress.wsu.edu

First printing 2017

Printed and bound in the United States of America on pH neutral, acid-free paper.

Library of Congress Cataloging-in-Publication Data

Names: Lovin, Hugh T., author.
Title: Complexity in a ditch : bringing water to the Idaho desert / Hugh T. Lovin ; introduction by Adam Sowards.
Description: Pullman, Washington : Washington State University Press, [2017] | Includes bibliographical references and index.
Identifiers: LCCN 2017022064 | ISBN 9780874223538 (alk. paper)
Subjects: LCSH: Irrigation--Idaho--History. | Water resources development--Idaho--History. | Water-supply, Agricultural--Idaho--History.
Classification: LCC HD1739.I2 L68 2017 | DDC 333.91/309796--dc23
LC record available at https://lccn.loc.gov/2017022064

WSU Press acknowledges the generous support of Jeffrey D. Lovin in making this book possible.

On the cover: Cottonwood flume south of Kimberly, Idaho (Milner Dam & Main Canal: Twin Falls Canal Company). *Bisbee Photo, William H. Eaton photographer, date unknown. Library of Congress 059465; original print located in the Twin Falls Library.*

Dedication

To Ida Carolyn Lovin, Hugh Lovin's wife of 58 years.

Contents

Illustrations

Maps

Publisher's Preface

In 1998, historian Hugh T. Lovin, recently retired from Boise State University, submitted a manuscript for consideration to WSU Press. Working with then-editor Keith Petersen—who would later become Idaho's State Historian—Lovin proposed gathering a collection of his articles on water in Idaho into a book he tentatively titled "The Irrigation West's Dreamers, Schemers, and Doers at the Snake River Plain." For myriad reasons—staff changes, the passage of time—the project never came to fruition.

Fast forward sixteen years to 2014, when WSU Press editor-in-chief Robert Clark received a phone call from Jeffrey Lovin, MD, who had found in his father's records a copy of the original proposal and was interested in resurrecting the project as a gift and tribute to his aging dad. Within a few days, Judy Austin received a call from Jeff, and quickly leapt on the case, tracking down copies of Hugh's articles, including those on which she worked with him as editor of *Idaho Yesterdays*. She generously shared with us her time, her knowledge of Hugh and his work, and her knowledge and understanding of Idaho history and the critical role of water in the Idaho story.

Early enthusiasm from Mark Fiege and Keith Petersen for Hugh Lovin's work, and incisive, eloquent reviews of the manuscript by Laura Woodworth-Ney and Adam M. Sowards, confirmed the value of the project. When Adam agreed to write an introduction placing the work in the larger context of water in the West, the project was fast-tracked for approval by the WSU Press editorial board.

Our sincere thanks to the following for their permission to reprint Hugh Lovin's articles: the Idaho State Historical Society for articles from *Idaho Yesterdays*; the *Journal of the Southwest* for articles in *Arizona and the West*; and *Pacific Northwest Quarterly* and *Agricultural History*. Our hope is that a new generation of scholars will benefit from the deep and meticulous work of Hugh T. Lovin.

Foreword

Hugh T. Lovin in the 1970s.
Courtesy Jeffrey Lovin

Hugh Lovin (1928–2014) grew up on a farm in southeast Idaho—a farm watered by irrigation, the only possible way to raise most crops on southern Idaho's arid land. The system used on his family's farm was created by the community of Marsh Creek farmers themselves, who built dams on the creek and channels to their farms. Lovin did not stay on the farm; he did his undergraduate work at Idaho State College (now University) in Pocatello, a master's at Washington State University (1956), and—while serving as an instructor in history at the University of Alaska's military branch at Elmendorf Air Force Base—his PhD in history at the University of Washington. In 1965 he returned to Idaho to join the history department at Boise State, where he taught into the mid-1990s.

Lovin did not begin writing about irrigation until 1978. His doctoral dissertation was on the Spanish Civil War, and his initial published research was on politics and labor. Not politics in general, politics to the left and, if you will, the anti-left: the farmer-labor movement, the IWW, the Red Scare and vigilantism during World War I, the Progressive movement, the Nonpartisan League. Most but not all of this work was on Idaho events; two were on Lyndon Johnson and the Subversive Activities Control Board, some on other states' experience, some more general.

The articles that Lovin wrote for the Idaho State Historical Society's journal, *Idaho Yesterdays* (which I edited from 1967 to 2002), and other journals were admirably researched and written, scholarly in the best sense of the word. And they are about people, not just about beliefs and tensions. In other words, they are accessible not only to fellow historians but to the general reader as well. I've always assumed that they reflect some of his own views, but we never talked politics in the years that I had the pleasure of working with him.

When Lovin began writing about irrigation (and, sometimes, irrigation politics), he again focused on people's experiences. Much of his work is about projects under the Carey Act, the 1894 law that established a complicated relationship among the federal government, states, private developers, and settlers. Lovin wrote about failures as well as successes, and he was intrigued by some of the quirkier examples of Carey Act projects. Politics and ideology are not the point here, though in some cases they may be a part of the story. Lovin wrote more generally about irrigation as well, including the opening chapter in this volume, "Dreamers, Schemers, and Doers of Idaho Irrigation." His essays are among the best work on a critical subject in Idaho history that needs much more scholarly treatment, making this collection especially valuable.

Lovin's two streams of research—on the development (and non-development) of irrigation projects and on more liberal influences in Idaho's politics—have led to both invaluable studies and good reading. They set an example for generations of students at Boise State. As his frequent editor, I learned an enormous amount about subjects in which I already had an interest. We shared a love of Idaho and its history—and we also shared a belief in the importance of its citizens' understanding of that history. Hugh Lovin's career as both teacher and writer was deeply committed to that belief. He was a warm and welcome colleague in helping Idahoans better understand themselves.

Judith Austin
Idaho State Historical Society, retired

Finding Complexity in a Ditch: Hugh T. Lovin and Idaho Irrigation History

ADAM M. SOWARDS

In history, the Snake River Plain has often been an obstacle. The storied nineteenth-century overlanders looked out from their wagons and saw desolation through most of the landscape that stretched nearly four hundred miles long and up to one hundred miles wide.[1] For decades, regional promoters saw plenty of arable land but it was much too far from water. Even modern observers, such as Hugh T. Lovin, whose work is collected in this book, characterized portions of the plain in less than inviting terms: "Mostly a hot, uninviting desert, this region was covered with straggling sagebrush and strewn with flinty outcroppings of volcanic rubble that impeded travel."[2] Yet today, if you fly over southern Idaho—or zoom in on Google Earth—you find huge bands and circles of green and, looking closely, see reservoirs storing the region's scarce water supply. The desert has been transformed. Early dreamers imagined five million acres (the delusional envisioned up to nine million) irrigated throughout the Snake River Plain. The first generation of irrigation projects, those completed before 1920, put two and a half million acres into production through 13,000 miles of ditches and canals flowing to 18,000 farms. All of this helped fuel the state's population growth fourfold in just two decades between 1900 and 1920.[3] Plenty of failures and hardship line the historical route to the present, but it is hard to argue against the green results of irrigation.

As the nineteenth century closed, the American West faced a challenge of accommodating a new, growing population that demanded more from nature, especially its rivers. So, civic boosters, technological enthusiasts, and government agents turned attention to ways of maximizing opportunity. Because past traditions seemed to be failing, or were at least inefficient, they adopted new practices and forged new policies to develop the West, yielding success, failure, and combinations of the two.[4] Then, much later, as the twentieth century wound down, new challenges came to accommodate ever more people and demands on a shrinking water supply, and government officials and environmental activists wondered if there might not be a better way. For instance, in 1977 President Jimmy Carter unveiled a so-called hit list of proposed water projects he planned to eliminate while Earth First! rabble-rousers proposed breaching dams.[5] It would surprise only the unobservant that writers and historians during the late twentieth century turned their attention to the century before to understand how and why the West built its hydraulic infrastructure and its irrigation communities as it did. Hugh Lovin exemplified this historical practice at that moment—looking backward from a time of uncertainty to investigate a similar moment of flux. He did not toil alone, in isolation, or from a blank slate. Other writers and scholars explored irrigated landscapes, examined water regimes across many locales, and built on earlier reclamation traditions. Situating this book and Lovin, then, requires attention to these contexts and Idaho itself.

Like the tobacco and wheat crops they harvested, American farmers grew the nation, a fact easy to forget today when suburban and urban populations far outpace their rural counterparts. The farms that nourished the young nation's economic and political roots sat in the humid eastern half of North America. Insufficient water rarely caused long-term problems for those agricultural communities. Thomas Jefferson, among others, viewed the nation's virtuous farmers as the republic's bedrock. Although the Jeffersonian foundation of American agriculture had problems—slavery being the most obvious—environmental limitations seemed minor and only occasionally inconvenient. The nineteenth-century myth of inexhaustibility also propelled the United States' citizens west, and they brought with them cultural expectations that farming would continue much as it had along the Atlantic seaboard and eastern river valleys. Public policy reinforced and powered this movement of people and products, auctioning off and giving away the public domain. But then, the American West threw up the obstacle of aridity, an environmental challenge that demanded readjustments to communities, law, and institutions. As one of nineteenth-century America's most celebrated explorers John Wesley Powell plainly put it, "the climate is so arid that agriculture is not successful without irrigation."[6]

People inhabited the American West from time immemorial, developing varied adaptive measures to survive in arid places. These included irrigation systems, such as the vast complexes among the Hohokam in what is now central Arizona or the flood irrigation methods used by various tribes in the Southwest. Others grew food along watercourses, adjusting seasonal mobility to ensure they were in the right place at the right time to plant, tend, and harvest crops. Droughts certainly occurred and severe ones forced indigenous groups to relocate. But thousands of years of successful adaptation demonstrated that civilizations could thrive in western North America.[7]

However, the Euro-Americans who began exploring the far West in the late eighteenth and early nineteenth centuries saw the arid landscape with a different set of cultural eyes and social practices. Some of the earliest, like Zebulon Pike, called it the Great American Desert, effectively discouraging resettlement by Euro-Americans. However, Latter-day Saints worked together to build communal irrigation systems, including the Great Feeder Canal in what became southeastern Idaho, that succeeded and inspired others, although the close-knit nature of Mormon communities and the power of the church proved difficult to emulate elsewhere. Eventually boosters and a new national mindset promoted and encouraged individuals and families to repopulate the West after dislodging and marginalizing Native Americans through political and military might.[8]

Legally, water became like property—something that could be bought, sold, or transferred—after the Spanish and Mexican communal traditions gave way to American imperialism and its governing institutions. Prior appropriation—usually abbreviated as "first in time, first in right"—guided most western territories and followed the practices originating in the West's mining lands. This legal practice set important precedents. It encouraged early use of water, as well as continuous use, for if farmers stopped using it their right would disappear. As opposed to riparian rights, which derived more closely from English common law, prior appropriation divorced water rights from the land, which allowed farmers to import water from off their property, often a prerequisite for irrigating western farms.[9]

The federal government tried to provide for an orderly process to get the public domain into private hands through laws like the 1862 Homestead Act. The first step in creating that order was a government survey. When government surveyor and polymath John Wesley Powell returned from his investigation of the interior West, he wrote his *Report on the Lands of the Arid Region*, which appeared in 1878. The *Report* issued would-be farmers and enthusiastic lawmakers a stern warning: "Many droughts will occur; many seasons in a long series will be fruitless; and it may be doubted whether, on the whole, agriculture will prove remunerative." In highlighting the challenges the region presented to American agriculture, the *Report* signaled a need to reshape land law and governance. For instance, in proposed legislation Powell included in his report, he allowed groups to organize irrigation districts to claim public lands with individual plots limited to 80 acres (instead of the 160 acres in the Homestead Act and larger acreages in other land laws). Many western lands simply were not useful when divided in surveyors' sections but needed to be organized around water availability. In general, Powell offered more cautious and less individualistic plans than most Americans preferred, and his conclusions and recommendations flew in the face of tradition and predilection; before too much time passed, he lost his job.[10]

Needless to say, the federal government did not follow Powell's visionary suggestions, but eventually it had to adjust its practices. For example, it incentivized individuals to develop irrigation with the Desert Land Act, passed in 1877, that gave more acreage to a farmer who promised to bring irrigation to it. This failed and attracted fraud, as people and companies sought ways to acquire as much land as possible even it if they could not farm it. The business of land in the West was big business indeed and attracted the unscrupulous and genuine investors alike.[11]

The existing land system proved not as productive as many thought it ought to be. As journalist Marc Reisner explained, "For the first time in their history, Americans had come up against a problem they could not begin to master with traditional American solutions—private capital, individual initiative, hard work—and yet the region confronting the problem happened to believe most fervently in such solutions." And so, the federal government tried something new with the Carey Act (1894), a notable shift in government investment. The act gave desert states up to one million acres of federal public domain provided that the state get the lands irrigated, using private corporations and investors who profited by selling the water to farmers on the lands. The legislation largely failed across the West. But Idaho proved the exception, developing almost the full million-acre allotment and sending Idaho leaders to Washington, DC, asking for more. This success, along with the concomitant difficulties and setbacks, furnished Lovin much historical fodder.[12]

At the time, a national campaign for reclamation poured over the land like flood irrigation covered fields. William Ellsworth Smythe led the charge, publicizing all good things that would come from building irrigation works across the West. Writing at a moment of national imperialist fervor—the country had just emerged victorious with colonial possessions after the Spanish-American War—Smythe envisioned a western, domestic colonialism as a better bet. Filling up the West—characterized as *The Conquest of Arid America*, as he titled his 1900 book—would produce prosperity and a democratic one at that. Irrigation was a miracle, one chapter explained, that promised democratic communities comprised of Jeffersonian small farms. In Idaho, Smythe saw great potential, because the state "has barely crossed the threshold of its vast possibilities." What Smythe saw in underpopulated arid spaces of the West, like Idaho's Snake River Plain, were the roots of a greater republic, places where irrigation would help tie people to the land and help propel a cooperative social and economic evolution that would fulfill America's potential. To advocates like Smythe, the transformation remained central to the *American* story, not just a *western* one.[13]

Tapping into this enthusiasm—and Idaho's exceptional success notwithstanding—reformers pushed for even greater federal assistance that they hoped might solve the shortcomings of the Carey Act, resulting in the National Reclamation Act, also known as the Newlands Act, of 1902.[14] This legislative program inserted the nation-state to a greater extent into efforts to promote and build irrigation systems in western territory. It would use proceeds from public land sales to seed a reclamation fund distributed for constructing dams and canals to expand the West's irrigated acres. Then: new homes, new farms, new crops, new money would spring forth from western deserts and sagebrush plains. With a limit of 160 acres on these federal projects, the law presupposed small-scale family farms, ever the American Jeffersonian hope, and showed how the reclamation campaign fulfilled ideological impulses as well as economic functions. Yet, by creating a new federal agency—the Reclamation Service, later promoted to the Bureau of Reclamation—and funding it through federal dollars, the act put the national government more firmly in the reclamation business.[15]

Although proceeds from successful irrigation projects were meant to repay the reclamation fund, results were mixed. After two decades, a scant 10 percent of the funds had been repaid and 60 percent of farmers were in default on their repayments. Ultimately the federal government forgave many projects' debts. Meanwhile, the Depression of the 1930s arrived and reclamation projects grew larger in response to the need for labor and the greater technical expertise engineers acquired. Irrigation history did not conclude with the Depression, but Lovin's interest remained firmly rooted in these initial decades.[16]

And for good reasons: those years were exciting times for Idaho. A chief engineer on the Twin Falls North Side Land and Water Company, E. B. Darlington, captured that zeitgeist in an essay,

"The Romance of a River," that appeared in 1920 in *Reclamation Record*. The engineer's romance grew out of an imagination as fertile as the Snake River Plain. Darlington shared a history lesson with irrigation engineers—"men of great vision and master designers"—at the center, who considered themselves "understudies of the Creator, delegated to bring forth upon the earth a better condition of life, a finer spirit of contentment, a higher state of development, and an advancement in human progress." The future beckoned for more of the same, but pitfalls lurked if farmers followed reckless plans. Darlington concluded that duty required the river be transformed into an agent "only of beneficence…for the greatest good of the greatest number." Darlington, who later served as the superintendent of the federal Minidoka Project, embodied the confidence of the era, a belief that experts could produce widespread benefits and "full utilization" of natural resources.[17]

Darlington's article showed how irrigation was always about fulfilling visions—of farmers, engineers, and politicians; in short, of anyone who hoped to transform desert spaces into productive farms. Historians like Hugh T. Lovin examined these stories, trying to show the ways these reclamation impulses served progress and development. But historians found many reasons not to share in Darlington's unequivocal enthusiasm.

In the mid-1980s, as Lovin researched and produced many of his studies, two major books on western irrigation appeared—one by a journalist and one by a historian—and each used the past to frame and explain then-present concerns. Both saw the West's current hydraulic regime as failing, a system that subverted both environmental and democratic ends. For the journalist Marc Reisner, watering the West came from a deluded biblical mission to make the desert bloom as a rose, an effort he characterized as "messianic." The book's epigraph, Percy Bysshe Shelley's elegiac (or is it prophetic?) sonnet "Ozymandias," evoked the inevitability of declining power in desert civilizations. Further, Reisner's story in *Cadillac Desert: The American West and Its Disappearing Water* abounded in bureaucratic rivalries between the U.S. Army Corps of Engineers and the Bureau of Reclamation competing for ever greater projects, rejecting economic and ecological rationality as a matter of course in the pursuit of administrative power. Dam failures, cost overruns, and corruption marked the West's history with reclamation, as surely as cowboys and Indians rode through 1950's westerns.[18]

The historian Donald Worster, on the other hand, connected the West to a long and global history, one rooted in despotism that grew out of past societies that sought to control nature—especially water. Controlling water, Worster maintained in *Rivers of Empire: Water, Aridity, and the Growth of the American West*, allowed ruling elites to control people in places like India, China, and the American West. Applying insights from thinkers like Karl Marx and Karl Wittfogel, Worster argued that rather than a place synonymous with freedom as it liked to think of itself, the American West "is increasingly a coercive, monolithic, and hierarchical system, ruled by a power elite based on the ownership of capital and expertise," best shown by the irrigation canal. To Worster, the West's hydraulic empire was a creation of the state melded with capitalism that wreaked unequivocal environmental and social havoc by serving an agribusiness immune to the people's interest.[19]

Both Worster's and Reisner's stories of the West show a certain imperialism at work. They see the West as a place colonized and manipulated by bureaucrats, engineers, politicians, and more, a region where nature itself was shackled to a statist-capitalist imperative. Consequently, the books are akin to medieval morality plays.[20] They are eloquently written, passionately argued, and, if not exactly caricatured, then perhaps they sell exceptions as more of the rule than is merited.[21] From that perspective, they might have become too influential. In these accounts, too, California and the Colorado River play outsized roles, exaggerating their history as indicative of the West writ large.[22] So, the history of irrigation in a place like Idaho might reveal other historical contours. This is why we need Lovin's work.

Building farms in Idaho's Snake River Plain was difficult and risky. Land needed clearing; water needed moving; pests needed removing; droughts needed avoiding. These factors and more had stalled agricultural expansion in the late nineteenth century before large-scale reclamation projects developed, and they would continue as ubiquitous challenges to Idaho farmers even when more investment and greater technological might arrived. Beyond environmental forces, aspiring reclaimers required steep capital investments and complex engineering works to transform a sagebrush-covered range into sugar beets, alfalfa, orchards, and, of course, potatoes. Money often ran out, even for scrupulous investors (and not all were). Meanwhile, gravity-fed canal systems were prone to failure and pumping systems required power, which made them more expensive. Laying out irrigation tracts required audacity in vision and comfort with risk against sometimes high odds. Against such obstacles, it's sometimes a wonder any irrigation systems got built.

So, in 1890 or 1920, what interested locals most was not grand theories about state power and despotism but getting canals to deliver water to fields so crops could grow and capital could flow. And no one accounted for this process in Idaho better than Lovin. What he showed in his many articles, some of the choicest pieces represented here, was no clear overriding thesis that might interpret all of Idaho's reclamation experience. After all, Idaho often fits poorly in sweeping historical generalizations, and reclamation history is no exception.[23] No one demonstrates this better than Lovin. His collective work succeeded where the morality plays failed in part because of scale. Often, his research focused on a single project, trying to account for its success or failure or some creative mix of the two—seemingly the most common outcome. Lovin developed these nuanced histories through painstaking research and attention to historical detail. That was often the criticism of other books: Worster's monolithic state was, well, too monolithic and did not reckon with the fragmented and localized nature of the nation a century ago.[24]

When Lovin searched for Idaho's history in an irrigation ditch, he found complexity. At one point, after years of research, Lovin characterized—typologized really—the main actors in Idaho reclamation efforts as "dreamers, schemers, and doers."[25] The labels functioned as a sort of shorthand for those who imagined a prospering plain irrigated by the Snake and its tributaries (dreamers), those who manipulated images and often others' money to promote this or that tract with an eye for profit (schemers), and those who buckled down and built the dams, canals, and farms that transformed the dry, fat bottom of the state (doers). The lines between types blended too easily, with shady developers promising easy riches and state agents promising water before it was available in close collaboration a century ago in a state hurrying to grow.

To a substantial degree, the state grew, as Lovin showed, through bringing irrigation to undeveloped landscapes. The process seemed straightforward: sagebrush (and rabbits) had to be removed; land had to be sold; and water had to be provided. But within those parameters much trouble might be made. Government officials—state and federal—overpromised water, leading to shortfalls. Financiers overpromised money, leading to bankruptcies. Engineers overpromised technical solutions, leading to system failures. At least sometimes. At other times, it worked. Investors like Frank Buhl and Peter Kimberly faithfully put together solid projects like the Twin Falls South Side project that irrigated nearly a quarter-million acres, a model under the Carey Act and envy of others.[26] Federal reclamation built its own projects, such as Minidoka and Boise, helping to create Idaho's landscape and build its economy. Lovin's work demonstrates how mammoth was the task of making the desert bloom as the rose, which promoters so easily promised.

To be sure, Lovin had blind spots or areas that may not have interested him. His histories are essentially about establishing projects and the political, financial, and technical requirements to do so. Cultural and social histories of irrigation might not have even occurred to him, although we now have searching analyses of some of the art and literature inspired by Idaho ditches, as well as personal reflections on water by Idaho authors.[27] Studies and accounts of the way women both promoted and experienced reclamation projects add a contour Lovin neglected.[28] The sticky and ubiquitous issue of indigenous water rights and how they intersect with irrigation regimes did not gain attention from Lovin, although we know the importance of tracking those relationships.[29] The labor required to build irrigation works and, then, to plant, tend, and harvest the crops that are the ultimate products of reengineered rivers never drew Lovin's focus. And scholars still have done little to see the land-labor nexus in Idaho fields and along Idaho rivers as they have in other locations.[30] Concerns about the environment—concerns that occupied other histories, especially as Lovin's career wound down—played a minimal role in Lovin's scholarship.[31] Had he carried his investigations further into the twentieth century, Lovin surely would have had to reckon with questions related to groundwater pumping.[32] Any scholar working on reclamation in Idaho today would be expected to consider at least some of these topics.

None of which is to minimize Lovin's achievements collected in the exemplars contained in this book. In the pages that follow we confront the work of a careful historian, a scholar who focused a career on understanding one of the most fundamental elements that built the state that paid his salary as a professor at Boise State University. He pursued no overt political agenda. Lovin evinced a partisanship neither for free enterprise nor federal reclamation projects; his was a partisanship dedicated to historical facts, as he found them. Mythology didn't blind him in the archives. The result is a body of work that stands up to the test of time and helps Idahoans understand how their irrigated landscape came to look so much as it does today.

Adam M. Sowards is professor of history and director of the Program in Pacific Northwest Studies at the University of Idaho. He is the author of several articles and books, including *The Environmental Justice: William O. Douglas and American Conservation*, and the editor most recently of *Idaho's Place: A New History of the Gem State*.

Notes

1. Peter G. Boag, "Overlanders and the Snake River Region: A Case Study of Popular Landscape Perception in the Early West," *Pacific Northwest Quarterly* 84, no. 4 (Fall 1993): 122–29.

2. Hugh T. Lovin, "A 'New West' Reclamation Tragedy: The Twin Falls-Oakley Project in Idaho, 1908–1931," *Arizona and the West* 20, no. 1 (Spring 1978), 5–6. Reprinted in this volume as Chapter 7.

3. Most figures are Lovin's and are repeated in many of the following chapters. The outlier number of nine million acres came from the Idaho State Bureau of Immigration and is from Lovin, "Water, Arid Land, and Visions of Advancement on the Snake River Plain," *Idaho Yesterdays* 35, no. 1 (Spring 1991), 7 and is the Epilogue in this volume. For the canal mileage and farms, see Mark Fiege, *Irrigated Eden: The Making of an Agricultural Landscape in the American West* (Seattle: University of Washington Press, 1999), 23.

4. Perhaps the best overview of the efforts to develop the West's natural resources is Charles F. Wilkinson, *Crossing the Next Meridian: Land, Water, and the Future of the West* (Washington, DC: Island Press, 1992).

5. On Carter, for instance, see Marc Reisner, *Cadillac Desert: The American West and Its Disappearing Water*, revised and updated (New York: Penguin, 1993), ch. 9; for Earth First! see Susan Zakin, *Coyotes and Town Dogs: Earth First! and the Environmental Movement* (New York: Viking, 1993).

6. William deBuys, ed., *Seeing Things Whole: The Essential John Wesley Powell* (Washington, DC: Island Press, 2001), 156. The interpretation offered of agricultural expansion and the challenges of western aridity is a standard one. A good account can be gleaned from Richard White, *"It's Your Misfortune and None of My Own": A New History of the American West* (Norman: University of Oklahoma Press, 1991), esp. chs. 6, 9, 15. A useful account of Jeffersonianism and its complicated reality is Roger G. Kennedy, *Mr. Jefferson's Lost Cause: Land, Farmers, Slavery, and the Louisiana Purchase* (New York: Oxford University Press, 2003).

7. Adaptive measures to aridity and drought are described in William deBuys, *A Great Aridness: Climate Change and the Future of the American Southwest* (New York: Oxford University Press, 2011), ch. 3; Norris Hundley Jr., *The Great Thirst: Californians and Water: A History*, revised ed. (Berkeley: University of California Press, 2001), ch. 1; Adam M. Sowards, *United States West Coast: An Environmental History* (Santa Barbara, CA: ABC-Clio, 2007), 74–78.

8. Although the main topic of the book is John Wesley Powell, Wallace Stegner accounts for the booster attitude imbued in the West in his classic, *Beyond the Hundredth Meridian: John Wesley Powell and the Second Opening of the West* (1953; reprint, New York: Penguin, 1992). One example that shows the displacement of Native peoples is Ned Blackhawk, *Violence Over the Land: Indians and Empires in the Early American West* (Cambridge: Harvard University Press, 2006).

9. The best places to begin with the legal elements include Hundley, *The Great Thirst*, chs. 2–3; Donald J. Pisani, *To Reclaim a Divided West: Water, Law, and Public Policy, 1848–1902* (Albuquerque: University of New Mexico Press, 1992), chs. 2–3; Wilkinson, *Crossing the Next Meridian*, ch. 6.

10. For context, see Stegner, *Beyond the Hundredth Meridian*; White, *"It's Your Misfortune,"* chs. 5–6. For Powell's plan, see deBuys, ed., *Seeing Things Whole*, 149–208; quotation from 158; plan specifics from 192.

11. The most thorough account of federal land law is Paul W. Gates, *History of Public Land Law Development* (Washington, DC: Government Printing Office, 1968), 399–401 for Desert Land Law and its failure.

12. Reisner, *Cadillac Desert*, quotation from 110. For Carey Act, see Donald Worster, *Rivers of Empire: Water, Aridity, and the Growth of the American West* (New York: Oxford University Press, 1985), 157. Lovin summarized the Carey Act in Idaho in Hugh T. Lovin, "The Carey Act in Idaho, 1895–1925: An Experiment in Free Enterprise Reclamation," *Pacific Northwest Quarterly* 78, no. 4 (October 1987): 122–33 (appearing as Chapter 4 in this volume).

13. William E. Smythe, *The Conquest of Arid America* (New York: Harper & Brothers, 1900), quotation from 184; Smythe divided his book into four parts, the first concerning "Continental Expansion at Home," which culminated in chapter five: "The Miracle of Irrigation."

14. It is important to note that in Smythe's imaginings the federal government played a muted role. But his work helped legitimize the larger reclamation project, thus building support for what became the Newlands Act of 1902. On Smythe, see Worster, *Rivers of Empire*, 118–25.

15. The Newlands Act is a central piece of reclamation history, and it plays a central role in Worster's argument. See *Rivers of Empire*, 156–69. Also see Pisani, *To Reclaim a Divided West*, 273–325; Reisner, *Cadillac Desert*, 111–19.

16. The default rate comes from Reisner, *Cadillac Desert*, 116.

17. E. B. Darlington, "The Romance of a River: A Past, Present and Future Survey of Irrigation in Southern Idaho," *Reclamation Record* 11 (March 1920): 122–23. Fiege analyzes Darlington in *Irrigated Eden*, esp. 23, 172, 177. Darlington's later position on the Minidoka Project comes from a letter appended to an Idaho Department of Water Resources decision found at www.idwr.idaho.gov/files/legal/orders/20130211_AFRD2-Final_Order_Re_Inst_to_Water_Dist_1_Watermaster.pdf.

18. Reisner, *Cadillac Desert*, front matter, 3, and passim.

19. Worster, *Rivers of Empire*, quotation from 7; see also chs. 1–2.

20. Perhaps this genre is inherent when writing about western water, for Wallace Stegner's biography of John Wesley Powell certainly frames the narrative around competing moral characters, too. See Stegner, *Beyond the Hundredth Meridian*.

21. Perhaps the most astute and persistent critic of these views is Pisani. For a concise version of his criticism, see Donald J. Pisani, *Water and American Government: The Reclamation Bureau, National Water Policy, and the West, 1902–1935* (Berkeley: University of California Press, 2002), 283–84.

22. Worster's index includes no entries for Idaho or the Snake River, for instance. Hundley argued an alternative interpretation for California, suggesting that it is not Worster's California focus that inherently created his argument; see *The Great Thirst*.

23. I make a similar point about Idaho as a poor fit in my introduction, "Idaho's Place: Reckoning with History," in *Idaho's Place: A New History of the Gem State*, ed. Adam M. Sowards (Seattle: University of Washington, 2014), 3–10.

24. No one has pushed Worster on these points harder than Donald J. Pisani. See, for instance, *To Reclaim a Divided West*, 331–33.

25. Hugh T. Lovin, "Dreamers, Schemers, and Doers of Idaho Irrigation," *Agricultural History* 76, no. 2 (Spring 2002): 232–43 [Chapter 1 in this volume].

26. Besides Lovin's work contained in this volume, see Pisani, *Water and American Government*, 66–77.

27. See Richard W. Etulain, "Shifting Currents: Cultural Expressions in Idaho," in *Idaho's Place: A New History of the Gem State*, ed. Adam M. Sowards (Seattle: University of Washington Press, 2014), for instance, 241–42 on Mary Hallock Foote's fiction and 247–48 for her art; Fiege, *Irrigated Eden*, 171–202, which examines myth and metaphor of Idaho's landscape; and Robert T. Hayashi's meditation on race and place in *Haunted by Water: A Journey through Race and Place in the American West* (Iowa City: University of Iowa Press, 2007), which includes both an analysis of Idaho's reclaimed landscape and his own experience of the larger landscape of the place with water as a key theme. Mary Clearman Blew, ed., *Written on Water: Essays on Idaho Rivers* (Moscow: University of Idaho Press, 2001).

28. For women and Idaho irrigation, see Annie Pike Greenwood, *We Sagebrush Folks* (New York: D. Appleton-Century, 1934); Susan H. Armitage, "'Too Little, Too Late': Annie Pike Greenwood, Failed Sagebrush Pioneer," in *Terra Pacifica: People and Place in Northwest States and Western Canada*, ed. Paul W. Hirt (Pullman: Washington State University Press, 1998); and Laura Woodworth-Ney, "Water, Culture, and Boosterism: Albin and Elizabeth DeMary and the Minidoka Reclamation Project, 1905–1920," in *The Bureau of Reclamation: History Essays from the Centennial Symposium* (Denver: U.S. Department of the Interior, Bureau of Reclamation, 2008), 385–405.

29. One example of a scholar working through these issues is Amy E. Canfield, "'These Lands are Worthless without Water': The Federal Government's Divided Loyalties in Irrigating the Fort Hall Indian Reservation, 1902–1920," *Pacific Northwest Quarterly* 105, no. 3 (Summer 2014): 122–35.

30. Again, Fiege's *Irrigated Eden*, 117–42 is an exception, although it essentially ignores the Latino presence that has grown into an important source of labor. Errol D. Jones, "Latinos in Idaho: Making Their Way in the Gem State," in *Idaho's Place: A New History of the Gem State*, ed. Adam M. Sowards (Seattle: University of Washington Press, 2014), 201–34 offers a useful overview.

31. The best account of irrigating Idaho primarily from an environmental perspective is Fiege's *Irrigated Eden.*

32. Zachary A. Smith, *Groundwater in the West* (San Diego: Academic Press, 1989), 97–103. Also, Adam M. Sowards and Brynn M. Lacabanne, "Instituting Water Research: The Water Resources Research Act (1964) and the Idaho Water Resources Research Institute," *Water History* (2017), doi:10.1007/s12685-016-0190-x, http://link.springer.com/article/10.1007/s12685-016-0190-x, demonstrates some of these concerns.

PART I

Promoting and Settling the Irrigation Frontier

Idaho's Snake River Plain.

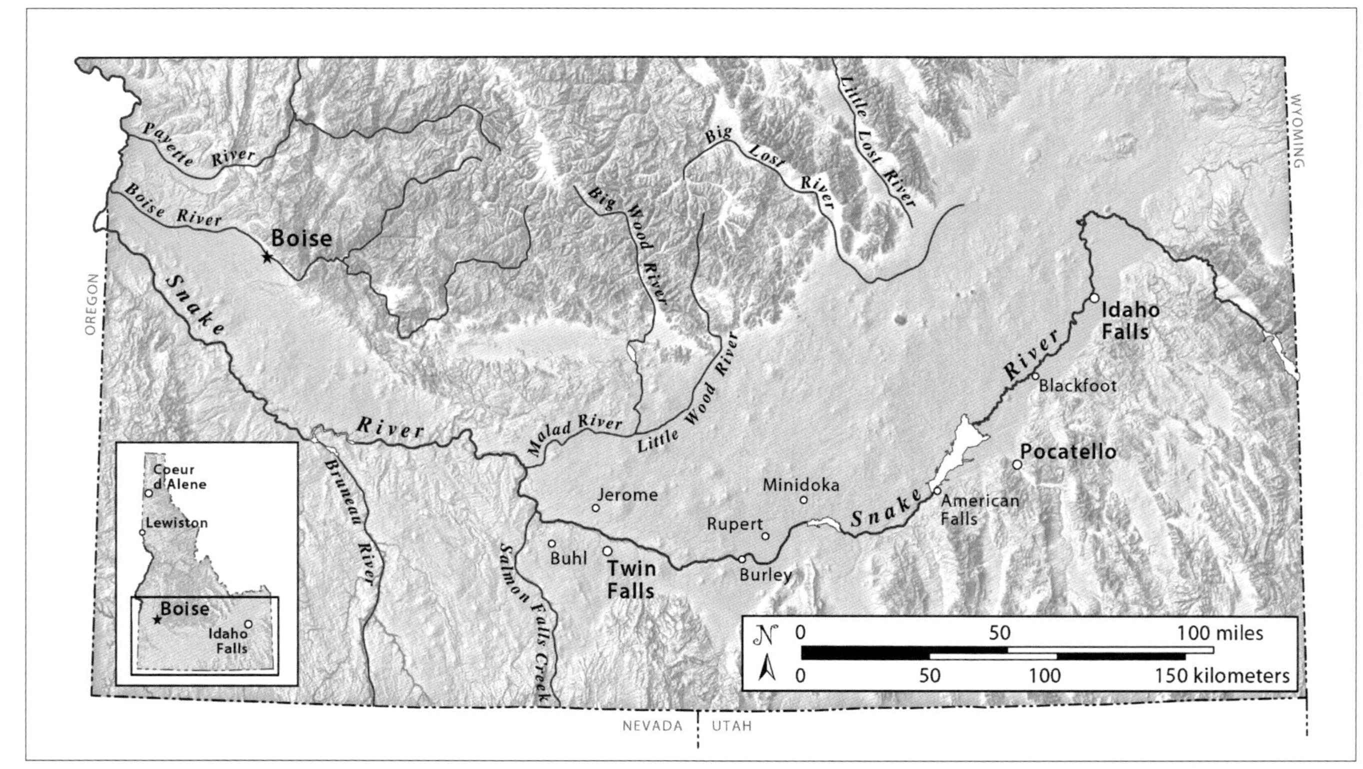

Map by Chelsea Feeney, www.cmcfeeney.com

CHAPTER 1

Dreamers, Schemers, and Doers of Idaho Irrigation

Frederick Newell, first director of the United States Reclamation Service, predicted in 1910 that western land reclamation projects would at first attract people who believed "roseate descriptions of irrigation," ignored warnings about privation ahead of them when they tried to carve farms from arid land, and soon fled from such adversity. Certain "second comer[s]," he said, might persevere until they overcame the odds against them succeeding, but the western irrigation frontier required a third wave of people who "may be regarded as the final locator."[1] Newell's axiom, in turn, was rooted so deeply in western irrigation lore that commonly two irrigators were expected to fail before a third one prevailed. Even at Idaho's Twin Falls South Side tract, where irrigation development happened about as expeditiously as anywhere in the state, reportedly the last of three "crops of settlers" finally achieved greater agricultural and social stability than any of the others.[2] But such arithmetical equations hinted only at the realities of land reclamation and scarcely addressed the real interplay of purity of social purpose, commercialism, and villainy among those who attempted arid-land reclamation across southern Idaho between 1880 and 1940. In Idaho, an army of visionaries, schemers, and doers labored but finally fell short of creating what a publicist claimed to be their mission on the Snake River Plain—to establish a "Garden of the Gods."[3]

Partly what these dreamers, schemers, and doers of Idaho irrigation did over the decades 1880–1940 was shaped by how irrigation was judged nationally. Outside Idaho, thinkers like William Ellsworth Smythe, publisher of *Irrigation Age* and author of *The Conquest of Arid America* (1899), proclaimed new magic-by-irrigation gospels during the 1890s, and such boosting for irrigation became even more infectious in commercial, publishing, and learned circles after 1900. The new theories so intrigued people like *Chicago Tribune* journalists that they accepted such ideas at face value; writing in the *North American Review*, another observer predicted the development of a new western empire comprised of 100 million irrigated acres. Agricultural scientists likewise touted the new thinking about irrigation. From Cornell University horticulturist Liberty Hyde Bailey and Iowa Agricultural College agronomist Perry Greeley Holden came glowing endorsements of irrigated farming. Writers for Bailey's agricultural cyclopedia placed the imprimatur of agricultural science on irrigation.[4] Hawking stocks and bonds of western irrigation companies, eastern and midwestern brokerages similarly emphasized notions of getting rich by irrigation farming.[5]

In part, the new irrigation gospels enjoyed acclaim because the new ideology embroidered the nation's old mythologies about the West. Such theories pictured irrigation as opening another roadway for Americans to travel on their journey to Eden. Such ideas about escaping from the wicked East to a blissful West remained the stuff from which human dreams were contrived long into the twentieth century. Furthermore, the irrigation gospels won public respect because of their high-minded arguments that in the case of the nation's social unfortunates, who had been shunted aside in the country's rush to industrialize, western irrigation offered them a second chance at social salvation. This notion of fostering irrigation for socially redemptive purposes gained even more support after President Theodore Roosevelt created the Commission on Country Life and placed Bailey at the helm. The commission handed its report to Roosevelt early in 1909. Its final words enlivened discussions that had already placed reaching Eden in the context of processes for reclaiming arid land in the West. The commission also called for arid-land reclamation under comfortable circumstances; hence, its proposals included schemes for bettering rural living with more material and cultural amenities. Moreover, the commission's work helped to inspire a back-to-the-land movement that, for the next two decades, encouraged city dwellers to flee from their urban abodes and find Eden at the end of their road.[6]

Ideas about creating one's own Eden on irrigated terrain piqued interest among eastern merchants who had tired of doldrums besetting Main Street commerce; but the prospects of making one's way to a western Eden appealed at least as mightily to plenty of people in the Midwest where, as in Iowa, too many people competed for the region's finite quantity of prairie land. So many midwesterners came to Idaho that early in the twentieth century, Iowa and Illinois clubs became commonplace; in many locales, the newcomers' Bible Belt mindset curried away the state's Old West subculture.

And the search for Eden in the arid West became even more a middle-class enterprise because lawyers, physicians, teachers, clergy, and clerical workers trekked westward in record numbers between 1900 and 1910. The group belonged, for the most part, to a generation imbued with the lofty ideals of socio-political progressivism for which Roosevelt was the ideology's best publicist, but the old sod seemed to supply this group neither enough professional opportunities nor sufficient chances to effect social uplifting. In 1906, Idaho's state immigration bureau reported shortages of craftsmen, hired hands for agricultural field work, and people willing to endure other types of hard labor despite the flocking of so many outsiders into the state. According to this bureau, "the class who perform common labor have not come in proportion to the small capitalist, the farmer, and the professional classes." For another decade, Idaho irrigation projects continued to absorb legions of middle-class people. At the federal Minidoka project, for instance, a special 1912 census of the entrymen revealed that many of society's more privileged folk persisted in swelling the ranks of those engaged in reclamation tract pioneering.[7]

For migrants to Idaho, dreamers like Smythe promised them extra dividends including rich water resources.[8] Those neophytes, unversed in irrigation realities, relied heavily on such representations because, in their reasoning, more water at their disposal equated proportionally to extra riches and greater well-being. But, on reaching Idaho, they learned at great expense to themselves that uncontrolled applications of water uprooted crops; worse yet, excessive watering induced so much alkali to accumulate on the surface of their land that no crops grew there.

Aside from encountering unfamiliar terrain and gravity irrigation technologies that puzzled them, the newcomers lamented other surprises as well. Firms like Clinton, Hurtt Company at Boise treated them fairly in dispensing land-locator services and information. But the newcomers more typically complained that the prophets of Eden had neglected to warn them of villains who met the unwary at the portals to Idaho and fleeced the immigrants. Echoing such complaints, a Colorado newspaper charged "land sharks of Idaho" with duping outsiders by the hundreds, and such newcomers bristled with indignation for good reasons. A Boise newspaper recounted the duplicity of promoters whose stock in trade consisted of luring eastern "suckers" to Idaho's land. Such enterprisers often styled themselves "Land Locators" but were simply real estate agents who collected liberal fees for steering settlers to arid land. The agents even knowingly placed their clients on land for which irrigation water would become available only after years of waiting for the moisture.[9]

Aside from running athwart such duplicity, Eden-seekers learned to their dismay that irrigation's prophets and reclamation tract promoters alike misrepresented the power and will of Idaho's state government to ensure that settlers received fair treatment. State officials actually offered them little protection in the dog-eat-dog capitalism by which even the best of Idaho's irrigation projects—save two federal projects—were financed, developed, and administered through their adolescence. But erroneous impressions persisted that Idaho took responsibility for shielding the settlers. A Boston businessman was not alone when he assumed that Idaho state authorities "aim[ed] to save the settler from disappointment, and the capitalist from investing in [irrigation] work[s] that will not bring expected results."[10]

Schemers elaborated on such myths by telling Eden-minded land seekers that entrymen enjoyed special state protections at projects that state authorities authorized and supervised under provisions of the federal Carey Act of August 18, 1894. Small wonder land seekers believed such a canard after encountering so much repetition of it. Like many others, the Kings Hill Extension Irrigation Company falsely advertised its tract as "Under State

Management." A newspaper claimed that Idaho guaranteed "a sufficiency of water for all irrigation purposes" at a different Carey Act project; a subordinate of the Idaho State engineer privately assured potential settlers of another project that the state "amply protected" them; and a United States Department of Agriculture official advised prospective settlers that Idaho mandated "a square deal" for all entrymen to Carey Act tracts.[11]

Settlers of Idaho irrigation projects found plenty about their new lot to deplore, but altogether the new irrigators experienced mixed results between 1900 and 1920. Seekers of Eden complained legitimately, saying that people readily lost their shirts in the twinkling of an eye; conversely, astute speculators garnered quick profits from investing in land and water rights on reclamation tracts. Furthermore, in the end, remaining on new irrigated farms despite multiple hardships significantly enriched irrigation tract farmers. In fact, after 1902, the Twin Falls South Side project became an exemplar for a decade of rapidly escalating farmland values that translated into personal wealth and security for landholders. However, one entryman remarked that such advances at Twin Falls came at the price of his spouse "wear[ing] herself out trying to keep the dust out of the house." At the South Side project, where conditions were no different than at nearly all of the state's new irrigation tracts, land reclaimers stripped the soil of its covering of sagebrush and native grasses at the outset. Wind then enjoyed free play with this barren earth until irrigation water finally dampened the soil sufficiently that it stayed in place.[12]

With so many Eden-seekers knocking at Idaho's doors, only irrigation enterprising on a grand scale could provide all of them land, water, and a fighting chance of prevailing. Happily for those seekers, they at first judged the federal government had favored them by launching the Payette-Boise and Minidoka projects within the state. But the federal reclamation tracts never satisfied all of the newcomers' demands for space and water, and they finally depended more frequently on private enterprisers to translate their dreams into reality by peddling enough irrigation company stocks and bonds to pay for large-scale irrigation works. In the beginning, such capital seemed to be within easy grasp. A Chicago newspaper predicted an outpouring of "millions of dollars" from this city to upgrade "the great irrigable section of the country." Nonetheless, such capitalistic processes were flawed, inefficient, and especially prone to failure at the hands of certain enterprisers. In one instance, the Mullins Canal and Reservoir Company pretended otherwise but actually placed only twenty acres under irrigation out of the company's 6,528-acre allocation.[13]

Bigger thinking about irrigation might produce results better than the Mullins Company's achievements, and even irrigation promoters in the same league as the Mullins group occasionally initiated large-scale reclamation projects despite jibing at them from the sidelines from Idaho old-timers. Idaho's oldsters called them "silk-hats" and despised their salesmanship of allegedly impractical irrigation schemes. Five silk-hats, for instance, each earned $5,000 for themselves by proposing to fill the Big Wood River valley with irrigated farms. Finally, they persuaded a Wall Street broker to arrange financing by outside investors of the project, and the business of putting water on this large acreage was at last underway.[14]

But depending on projects of large dimensions launched under such conditions could never suffice. Such projects developed too slowly to accommodate those Eden-seekers who could procure no place for themselves on the state's two federal irrigation projects. Clearly the times called for visionaries who possessed unassailable faith in irrigation endeavors, extolled large-scale irrigation technologies relentlessly, and overcame obstacles that irrigation's eastern prophets never mentioned. Furthermore, the chances of ever making fast progress along the road to Eden had shifted to the shoulders of free enterprise. Idaho irrigation's new visionaries needed charm, persistence, and guile to lure to the state enough money and an array of doers capable of creating large-scale irrigation projects aplenty. In particular, the new doers had to build and master the art of operating systems of dams across rivers, lengthy mainline canals, and thousands of waterways branching from the principal arterials.

Idaho's irrigation visionaries, albeit a parochial group of schemers, emerged early in the twentieth century and eventually claimed that the group had superintended the launching of a new hydraulic agricultural order in the state. In this group, Ira Burton Perrine stood out. After emigrating to Idaho in 1883, Perrine tried out mining and the stagecoach business before deciding that agriculture was a better bet. He established his irrigated "ranch" at Blue Lakes in the Snake River canyon in south central Idaho. At the same time, his vision but not his dollars kept the prospectively rich Twin Falls South Side tract before the notice of financiers until two doers—Pennsylvania ironmakers Frank Buhl and Peter Kimberly—agreed to complete this irrigation project. Kimberly soon died, but Buhl persevered and, in the end, rued his involvement at Twin Falls chiefly because his profits were not larger.[15]

The South Side project was scarcely underway when Perrine proposed grand-scale development of potentially irrigable land stretching from the banks of the Snake River, opposite the South Side venture, for about twenty miles to the north. He cajoled Pennsylvania banker William Speer Kuhn and his brother, James Speer Kuhn, until they agreed to provide an irrigation system for the Twin Falls North Side tract. Perrine subsequently insisted the vast Mountain Home plateau succumb to irrigation. He likewise called for irrigated farming instead of emptiness on the Bruneau plateau that was situated west of the South Side project; and, as land reclamation developed, Bruneau plateau irrigation stayed barely beyond Perrine's fingertips until the 1940s. During the same time, he advocated transmontane diversion of water from Yellowstone Lake (the largest of several lakes in Yellowstone National Park), to ensure fuller irrigation of the Snake River Plain. In 1919, he sketched his grandest single proposal for new irrigation—Perrine's so-called Franklin K. Lane project containing one million acres.[16]

As for the Kuhn brothers, who had listened to Perrine's preaching, a journalist stated, "no [commercial] proposition" was "too big to be considered." In the end, they launched three irrigation projects in Idaho that altogether contained nearly 500,000 acres at the inception; they created a fourth irrigation venture in California; and they dabbled in Idaho's budding railroad and hydroelectric power industries.[17]

Aside from the Kuhns, the first doers of large-scale irrigation in Idaho included several other entrepreneurs whose commercial designs nearly matched in size the grandeur of Perrine's visions for irrigation development. But many of their irrigation projects failed to prosper in the short run, and the big one in the Big Lost River valley collapsed. Even the Kuhns' irrigation empire toppled in 1913.[18]

The new doers of irrigation included enterprisers who exhibited less ambition than the Kuhns yet scarcely ever tackled the irrigation of so few as 50,000 acres. But these doers classed themselves in the Kuhns' league and carefully avoided the irrigation frontier's "grafters." Edwin Meredith—an Iowa publisher, subsequently governor of Iowa and later United States secretary of agriculture—insisted that he belonged to the group that reclaimed Idaho's arid land in good faith and practiced a type of capitalism that sold irrigation services in fair measure. Meredith even promised "to pull everybody through without loss" despite signs of abject failure at his West End Twin Falls irrigation project. Eventually, Meredith's project floundered. Many others were similarly unsuccessful despite the efforts of schemers to save them from abandonment. Although Governor Moses Alexander referred to one of these projects in particular, he explained what happened to many: the project "has had the mumps, measles and all the other childhood diseases," after which time its development remained at a standstill and only escaped "death and destruction by being well-nourished" verbally by its promoters.[19]

A second wave of doers invoked the likes of Buhl and the Kuhns during the 1910s and 1920s; like its predecessors, the group committed itself to satisfying Eden-seeker wants for land and water and to building a hydraulic agricultural order based on large-scale irrigation. However, the new group tried to avoid the errors of its predecessors who had underestimated the costs and non-monetary difficulties of reclaiming arid land on a grand scale. The work of the new doers, therefore, entailed patching the wounds of an old irrigation order, finding remedies to the mistakes of the first doers, and keeping alive the visions of crusaders such as Perrine while modifying old ideas to fit the economic, political, and environmental realities of the newer times. In trying to reshape irrigation in Idaho, the new doers depended heavily on counsel from engineers such as Andrew Jackson Wiley and Warren Swendsen. Wiley and Swendsen had honed their irrigation science in the school of hard knocks while serving as advisers to irrigation's first doers and sharing responsibility for shortcomings in many of the state's biggest irrigation projects.

The work of Russell Easton Shepherd exemplified the more enlightened approaches of the new crop of doers to resolving the difficulties of Idaho's irrigation tracts. A Minnesota lawyer who relocated to Montana in 1906, Shepherd came to Idaho about a decade later to represent bondholders of the Twin Falls North Side Land and Water Company. The bondholders had paid for developing the Twin Falls North Side project to its imperfect condition; their investment of $4 million was imperiled by the collapse of the Kuhn brothers' empire in 1913. Shepherd decided that the bondholders could recover few dollars. But he also concluded that guarding bondholder interests, as well as saving North Side settlers from losing their land and improvements on their farms, depended on somehow nursing the irrigation project to better health. Shepherd's new preoccupations ranged from fashioning solutions to the project's water shortages and financial travail to dealing with North Side agriculture's sociological ills. Shepherd could see that the North Side project was not an island; instead, its destiny intertwined with irrigation developments across the Snake River Plain. Consequently, he plunged deeply into the state's irrigation affairs to make regional improvements such as building a new system of reservoirs in the Snake River basin. He carried irrigation's banners from leadership positions that he accumulated in canal companies, irrigation districts, land banks, intermediate credit banks, chambers of commerce, New Deal agencies during the 1930s, and political parties.[20] Until his death in 1944, Shepherd continued crusading to find ways of bettering irrigation in Idaho after the results had belied the dreams.

This second string of doers founded few new projects but, at their best, nursed older irrigation tracts back to health. They frequently improved conditions at such places by reducing projects to a size that was commensurate with the water resources actually at hand. Such practices, however, gave water to certain Eden-seekers at the expense of others, and in places such as Churchill and Golden Valley in Cassia County, people were summarily cast aside and inevitably "lived and died for lack of water." For places like Taber and Kimama, irrigation's newest doers possessed no magic by which to relieve the miseries of people who had attempted to dry-farm desert land in the absence of reservoirs. In other instances, the ministrations of new doers helped farmers to gain more water but left the same farmers as vulnerable as ever to weather cycles and subsisting on variable streams of water from year to year. The commentary of a Wyoming farmer, in 1960, about his own circumstances near the headwaters of the Snake River just as accurately described the vulnerability of many Big Wood River, Big Lost River, and Salmon Falls Creek farmers after 1920. This farmer lamented "three years of good water supply and three years of inadequate or three years of medium." The consequence was a "boom-bust economy" in which a farmer "begins to get out of hock to the bank and then bang, he's hit with three years of inadequate water supply, doesn't raise anything, and is in it up to the hilt again."[21]

Architects and builders of grand-scale irrigation in Idaho often disappointed settlers and proved that irrigation's prophets had mistakenly pictured the road to Eden as short and straight. Nonetheless, irrigation's successive waves of doers accomplished more than their critics acknowledged. They had, at the least, demonstrated that irrigation was not inevitably a false idol for Eden-seekers. By Swendsen's estimates, Idaho farmers cultivated about 850,000 acres at Carey Act projects in 1924. In the same year, about 200,000 people lived within the confines of state government-sanctioned irrigation districts containing 1,723,673 acres. They irrigated the better part of this domain.[22]

But, in 1940, even the most fortunate of irrigationists also had good reasons to believe that the road to Eden stretched so far ahead and meandered so tortuously that reaching the end of it might require another fifty years. Each irrigation project had its own set of difficulties to surmount. Judy Austin's *Idaho Yesterdays* journal and Mark Fiege's *Irrigated Eden* (1999) analyze many of those problematic conditions. At the end of the twentieth century,

Idaho contained plenty of potentially irrigable land whereas its water resources were so finite that some of this land begged in vain for water. Given such conditions, the state's dreamers, schemers, and doers fell short of the mark that they had set for themselves. Clearly, the road to Eden in Idaho still meandered beyond the farthest horizon of human foresight; maybe all of the state's irrigable terrain can never be watered satisfactorily.

This article originally appeared in *Agricultural History* 76, no. 2 (2002). Reprinted with permission.

Notes

1. F. H. Newell, "Progress in Reclamation of Arid Lands in the Western United States," *Annual Report to the Board of Regents, 1910* (Washington, DC: Smithsonian Institution, 1910), 171–72.

2. Roy E. Huffman, *Irrigation Development and Public Water Policy* (New York: Ronald Press, 1953), 62; Alfred R. Golzé, *Reclamation in the United States* (Caldwell, ID: Caxton Printers, 1961), 374.

3. *Evening Capital News* (Boise), February 18, 1912, 3.

4. Donald J. Pisani, *To Reclaim a Divided West: Water, Law, and Public Policy, 1848-1902* (Albuquerque: University of New Mexico Press, 1992), 249–51, 285–94; *The Register and Leader* (Des Moines, IA), October 18, 1908, 10; *Shoshone Journal* (ID), September 17, 1909, 4; Donald Worster, "Freedom and Want: The Western Paradox of Aridity," *Halycon* 14, no. 1 (1992): 27; L. H. Bailey, ed., *Cyclopedia of American Agriculture: A Popular Survey of Agricultural Conditions, Practices and Ideals in the United States and Canada*, 4 vols. (New York: Macmillan, 1907–1909). See especially R. H. Hess, "Socio-economic Aspects of Irrigation," 4:168–70, and for materials on Idaho agriculture, scattered articles in volumes 1 and 4.

5. See, for example, *Irrigation Bonds Based on the World's Greatest Industry* (Chicago: Trowbridge and Niver, 1909), copy in Romaine Trade Catalog Collection, Davidson Library, University of California, Santa Barbara.

6. Robert G. Athearn, *The Mythic West in Twentieth-Century America* (Lawrence: University Press of Kansas, 1986), 31–33; Richard White, *"It's Your Misfortune and None of My Own": A New History of the American West* (Norman: University of Oklahoma Press, 1991), 405; Stanford J. Layton, *To No Privileged Class: The Rationalization of Homesteading and Rural Life in the Early Twentieth-Century American West* (Provo: Brigham Young University, Charles Redd Center for Western Studies, 1988), 5–19, 37–59.

7. *The State of Idaho: Official Report of the Bureau of Immigration, Labor and Statistics, 1905–1906* (n.p., [1906]), 107–108; "Report of Operation and Maintenance, 1912 [on Minidoka Project]" 68, Bureau of Reclamation Records, RG115, National Archives (hereafter NARG115). For accounts of how the newcomers fared on Idaho reclamation projects, see Richard Lowitt, "Irrigation Agriculture in Idaho as Seen by Henry A. Wallace in 1909," *Idaho Yesterdays* 35 (Spring 1991): 19–25; Richard Lowitt and Judith Fabry, eds., *Henry A.*

Wallace's Irrigation Frontier: On the Trail of the Cornbelt Farmer (Norman: University of Oklahoma Press, 1991), 118–68.

8. William E. Smythe, *The Conquest of Arid America* (1899; reprint, Seattle: University of Washington Press, 1969), 186.

9. *Denver Post*, n.d., as cited in *Rupert (ID) Pioneer-Record*, December 22, 1910, 1; *Evening Capital News*, March 30, 1906, 1; "[Boise Project] History, 1902–11," 17, Bureau of Reclamation, NARG115.

10. Charles S. Miller to Wayne Darlington, September 19, 1904, Idaho Reclamation Records, Collection AR-20, Idaho State Archives, Boise.

11. Randall R. Howard, "Irrigation Frauds in Ten States," *Technical World Magazine* 17 (July 1912): 505; *Twin Falls (ID) News*, January 6, 1905, 4; typescript copy of paid newspaper advertising, n.d. [1908], Kings Hill Extension Irrigation Company Papers, Idaho State Historical Society, Boise; *Shoshone Journal*, October 9, 1909, 1; Herbert Wing to A. B. Gilbert, December 2, 1905, and R. P. Teele to James Stephenson Jr., May 25, 1909, Idaho Reclamation Records, Idaho State Archives.

12. *Twin Falls News*, February 23, 1906, 1; Ernest G. Eagleson to Charles Addison Beach, April 5, 1907, Ernest G. Eagleson Papers, Idaho State Historical Society; Addison T. Smith to W. B. Heyburn, July 14, 1905, Addison T. Smith Papers, Idaho State Historical Society; also see Mark Fiege, *Irrigated Eden: The Making of an Agricultural Landscape in the American West* (Seattle: University of Washington Press, 1999), 11–80.

13. *Chicago Daily Journal*, n.d., reprinted in *Burley (ID) Bulletin*, March 16, 1906, clipping in Bureau of Reclamation, NARG 115; James Stephenson Jr., "Sixth Biennial Report of the State Engineer to the Governor of Idaho, 1905–1906," 24; and Stephenson to F. R. Gooding, March 18, 1907, Frank R. Gooding Papers, Idaho State Archives.

14. *Idaho Daily Statesman* (Boise), November 11, 1905, clipping in Bureau of Reclamation, NARG115.

15. Merrill D. Beal and Merle W. Wells, *History of Idaho* (New York: Lewis Historical Publishing, 1959) 2:139–140; H. J. Kingsbury, *Bucking the Tide* (New York: Ganis and Harris, 1949), 45–47; O. A. Kelker, "The Story of I. B. Perrine," *Twin Falls (ID) Times-News*, June 27, 1971, A-4, A-5; James A. Connelly, "Frank Buhl's 'Other' Home," *Herald* (Sharon, PA), April 22, 1976; "Frank Henry Buhl," *National Cyclopedia of American Biography* (Ann Arbor: University Microfilms, 1967), 24:433; Joseph Riesenman Jr., *History of Western Pennsylvania* (New York: Lewis Historical Publishing, 1943), 3:93–95.

16. "Irrigating the Country between Boise and Mountain Home," n.d., 1, David W. Davis Papers, Idaho State Archives; Mikel H. Williams, *The History of Development and Current Status of the Carey Act in Idaho* (Boise: Department of Reclamation, 1970), 28–30; F. E. Weymouth to D. F. McGee, November 28, 1919, Bureau of Reclamation, RG 115, File 201-I, National Archives Branch Depository, Denver, CO; "Minidoka Project History, 1919," 51, 54–64, Bureau of Reclamation, RG 115; "Franklin K. Lane Reclamation Project [of] One Million Acres…," n.d., William E. Borah Papers, Manuscript Division, Library of Congress; "Idaho Bruneau Prospectus and Suggestions," April 2, 1934, C. Ben Ross Papers, Idaho State Archives.

17. *Rupert Pioneer-Record*, February 25, 1909, 4; Williams, *Carey Act*, 72; Beal and Wells, *History of Idaho*, 2:145–47; *New California* (New York: A. Mestre, 1910), copy at New York City Public Library; *Sacramento Valley Irrigation Company: The Kuhn California Project* (n.p.: J. S. and W. S. Kuhn, n.d.), copy at Bancroft Library, University of California, Berkeley.

18. *New York Times*, July 8, 1913, 1, 2.

19. E. T. Meredith to Earl Sheets, November 11, 1913 (copy) and Moses Alexander to Senator William H. King, December 13, 1917, Moses Alexander Papers, Idaho State Archives.

20. Minutes of the State Board of Land Commissioners, September 10, 1919, Davis Papers; Byron Defenbach, *Idaho, The Place and Its People: A History of the Gem State from Prehistoric to Present Days* (Chicago: American Historical Society, 1933), 2:185–87; and for some of Shepherd's thinking about Idaho irrigation matters, see R. R. Shepherd, "The Financing of Irrigation Developments by Private Capital," *Transactions of the American Society of Civil Engineers* 90 (June 1927): 710–29; R. R. Shepherd, "Reclamation in Idaho," *New West Magazine* 11 (November 1920): 565–67.

21. *South Idaho Press* (Burley), April 17, 1972 and June 1, 1973, clippings in Al Dawson Scrapbook (copy), Idaho State Historical Society; Gerhard Riedesel, ed., *Arid Acres: A History of the Kimama-Minidoka Homesteaders, 1912 to 1932, By People Who Were There* (Pullman, WA: Gerhard Riedesel, 1969); *Upper Snake River Basin: Wyoming–Idaho–Utah–Nevada–Oregon* (Boise, Idaho: U. S. Department of the Interior, Bureau of Reclamation Region 1 and Corps of Engineers, U. S. Army Engineer District, 1961), 2:18.

22. W. G. Swendsen to C. C. Moore, July 20, and August 20 and 16, 1924, Charles C. Moore Papers, Idaho State Archives.

Grubbing sagebrush on the Ferguson Fruit and Land Company, 22 miles west of Twin Falls. *Twin Falls Public Library, Clarence E. Bisbee Collection, 172*

CHAPTER 2

Sage, Jacks, and Snake Plain Pioneers

Crescent-shaped and bisected by the Snake River, the Snake River Plain reaches over three hundred miles westward from the mountains that divide Idaho and Wyoming to the Idaho-Oregon border. High mountain ranges, seventy-five to one hundred miles apart, mark the plateau's north and south perimeters. The plain's topography is spectacular as well as dimensionally grand, for the river long ago carved deep canyons, and elsewhere this awesome region contains "flat bottom lands sloping with the [Snake] river to high plateaus falling back, terrace above terrace, to the foothills on either side."[1] "A region awful in its aloofness and inexplicable in its calm," novelist Vardis Fisher wrote in 1937, the plain has defied man to conquer more than a small portion of that "rolling mass of loneliness and waste, with the integrity of granite and the changelessness of time." Today much of it remains, in Fisher's words, an "empire of aridity and stone."[2]

Through nearly all of the nineteenth century, few tried to make the plain habitable; the evidence against such undertakings seemed conclusive. Viewing the plain from hills looming above Fort Hall, John C. Fremont wrote in 1843: "Covered as far as could be seen with artemisia [sagebrush], the dark and ugly appearance of this plain obtained for it the name of the Sage Desert."[3] An 1846 Oregon Trail guidebook described the area as largely a "world of waste and wrack"; an 1839 traveler reported not a single acre west of Fort Hall suitable for "grains or vegetables"; equally uncomplimentarily, Captain Wilson Price Hunt characterized the region a "dreary desert of sand and gravel."[4] More widely read and influential, Washington Irving powerfully condemned the Snake River Plain as a place to settle. His account of the Hunt expedition, for instance, stated:

> It is a land where no man permanently resides: a vast, uninhabited solitude, with precipitous cliffs and yawning ravines, looking like

> the ruins of the world; vast tracts that must ever defy cultivation and interpose dreary and thirsty wilds between the habitations of man.[5]

When venturers ignored these dire warnings, they often regretted their rashness. Mary Hallock Foote, a resident of the Boise Valley in the 1880s and 1890s, concluded that life there became bearable only when one resignedly overlooked the area's multiple shortcomings. Another, less malleable, immigrant decried her desert farmstead near Hazelton, surrounded as it was by "miles and miles of wilderness, and not a sign of habitation; no tree, no green, only the pungent sagebrush. And everywhere leaping jack-rabbits."[6]

But tarnished reputation and continued negativism about the Snake River Plain finally no longer deterred; instead, thousands scurried there when federal and state governments and private entrepreneurs opened several dozen reclamation tracts between 1897 and 1917. The newcomers came on the advice of publicists, journalists, reclamation-tract salesmen, railroads, and best-seller novelists, all of whom preached that "honest fortunes" were "going to waste" on the long-maligned Idaho deserts.[7] At the height of the frenzy, a Colorado newspaper portrayed South Idaho as "The Land of Opportunity"; another promised that an eighty-acre "homestead" provided the certainty of reigning "farmer king, independent and happy"; and Filer, Idaho, promoters advertised their area as the "Center of the Modern Garden of Eden."[8] An amateur bard said of the Twin Falls reclamation tract:

> Oh! Come to the land of lotus and honey
> Where pleasures are plenty and everything money.[9]

The newcomers sharply boosted Idaho's population, which rose from about 100,000 at the turn of the century to 413,866 in 1920. South Idaho counties received almost 88 percent of the influx.[10]

The new farmers, a county agricultural agent observed, ranged from "struggling homesteader" to "retired capitalist." Many of them "city people," a significant number were "tired-out professional[s] and business-men" who, grasping for a last chance at the proverbial pot of gold, had every reason to hope that prophets had not misrepresented the "New West's" opportunities.[11] In their haste to create personal Edens, most incoming agrarians shunned immediate comforts, first erecting crude housing on their domains. Tents or slightly more sophisticated canvas shelters could suffice in the southwestern counties where the climate was mild, but more commonly settlers built diminutive "prove-up" dwellings from tar paper over which they fashioned a wooden roof. On the Twin Falls South Side tract, such materials were obtainable for $125 to $150.[12] Playful settlers, unaware of their upcoming ordeals, placed signs on their homes proclaiming their unbounded optimism. "Paradise Avenue," one Twin Falls North Side hopeful christened his modest abode. Others, as if they sensed that these crude shelters might be their residences as long as they wrestled with the desert, used labels such as "Brimstone Bungalow."[13] However, those who expressed such negative outlooks incurred risks, for they brought suspicion upon themselves from reclamation-tract promoters and local civic boosters. The latter groups hated "knockers" and "croakers," and persistent pessimists might even be advised, as happened at Rupert, that a "croaker" was "a wart, a barnacle, and you had better sell out and move away. We don't need you and don't want you. Skiddoo!"[14]

Roofs above their heads, pioneers scanned their acres for renewed proof of fertility below the sagebrush that they must next remove. The State Engineer reported little difference in the richness of various Snake River Plain soils. However, if the sage was tall and bushy *artemisia tridentata* and their acres were free from low growths of greasewood, settlers exulted; local folklore and many experts equated the size and density of sage growth with soil fertility.[15] One expert, hired by the Twin Falls Land and Water Company to evaluate lands situated between the Salmon and Bruneau Rivers, classified the area "first class arid land." He reported:

> Those lands support at the present time a healthy growth of sagebrush and where the soil is deep and light in character, the growth is very rank. In one such place, we noticed a single bush which stood about a foot in height above a man's head on horseback. This is extra-ordinary. Both scientist and layman agree that sage-

> brush growth is an index of the general character of the land. No greasewood growth, indicative of excessive presence of injurious alkaline salts, was noticed on any of the lands classified as suitable for irrigation.[16]

Removal of the sage began immediately in order to reap the land's advertised riches sooner, destroy sage growths that harbored ticks whose bites could produce spotted fever, and, most important, conform to federal reclamation laws requiring settlers to cultivate at least one eighth of their holdings. One group, the affluent land speculators, simply employed hands enough to remove the sage from one eighth of the acreage, a process costing them $3 to $4 per acre, and then plant and irrigate a grain crop one time. Often they spent only thirty days meeting the General Land Office's requirements and, a Jerome newspaperman reported, then "took the train the same afternoon" after filing their proofs and "have not been seen since." Subsequently they paid required water and canal maintenance fees, leaving their land uncultivated until selling it.[17]

No easy roads lay ahead for the other settlers, although Boosters tried to lighten their burdens with humor. One joshed them that the "sagebrush sobbed as it fell before the sturdy blows of the man with the grubbing hoe."[18] Many had come to Idaho after ignoring responsible counsel that reclamation tracts were not "poor man's country" and a newcomer needed "capital and that in cash, otherwise he may find himself in embarrassed circumstances before he is well settled."[19] Consequently, they had to clear the sage themselves and many, lacking horses and implements, had no choice but to wield a grubbing mattock—with which, optimists advised, a single settler could clear one acre per day provided plenty of "elbow grease" was expended. Doggedly these pioneers grubbed the sage, laboriously piled it, and burned the uprooted brush in the evenings, keeping enough for domestic cooking and heating unless the settler could afford coal, and using the brush extensively for rip-rapping in canals and irrigation ditches.[20] Hence, only slowly did the sage succumb to the grubbers in many areas. A visitor to the Twin Falls North Side tract recalled that in 1910 still "a lot of gray, lonesome looking sagebrush" stood everywhere. Here and there, "little box shacks [were] stuck around over the country, and there were little patches where the sagebrush had been pulled off [out] and some crops planted."[21]

Clearers of sage encountered unpleasant surprises, for they discovered that the sage growths frequently concealed bounteous quantities of rocks. An angry King Hill project landowner complained that his acres were "absolutely good for nothing, except perhaps a very crude rock quarry." Denied a refund of his monies by irrigation tract promoters, he retorted that the promoters "should be wearing [penitentiary] stripes, in which event Society could heave a sigh of relief."[22] Removing tons of unwanted rocks joined grubbing and burning sage as recurring settler chores. "If any bunch of men ever worked," a veteran of this struggle with the desert later reminisced, "it was those poor devils."[23]

Settlers looked for less painful means of clearing their land. Instead of grubbing and piling sage to be burned, some fired fields of sage late in summer when ground vegetation had become "dry as tinder." Large areas could be cleared in that manner—but only at the risk of fires spreading beyond intended areas. In one of many instances of out-of-control fires, a fire near Burley threatened to destroy a dozen farmsteads and caused $3,000 in property losses before it was controlled.[24] Mechanical methods could provide another alternative, and necessity eventually became the agrarians' mother of invention. Farmers resorted widely to a "railing brush" method in which a metal railroad rail or similar object, the sharp edge of it placed next to the earth, was dragged over the sage to break off or smash the plants. Experimentation showed that the best results were obtained during cold weather; then the sage was brittle and snapped off more readily. Later Twin Falls-Salmon River tract settlers devised a frost grubber—a steel blade at an angle on a long beam usually referred to as a "sweep." The sweep engaged the sage, holding it until the sharp blade severed the plant at ground level. Laborious hand-grubbing, too, finally could be avoided after tinkering settlers stumbled across an important modification of their furrow-turning plows. When properly altered, the plows cut the sage at whatever depth the plows were set, leaving the plowed-up brush to be gathered and burned. When they could afford the implements, settlers also turned to heavy disk plows because the devices not only uprooted most of the sage but chopped sage, sage roots, and the stubborn sod beneath the sage in one operation.[25]

Rocks and sage removed, settlers hastily planted crops as soon as the first trickles of irrigation water became available. But uprooting the sage, a growth that most had come to loath, and removing rocks made more uneven the natural contours of land that was unready for planting in the first place without considerable smoothing. In their eagerness for immediate incomes, settlers ignored the requirement of level land with dire results. Writing in the *Hagerman Valley Sun*, Patrick McCaffrey reported:

> In traveling over the great [Twin Falls] North Side Tract one notices with pity the awful attempts of most eastern settlers to perform the simple trick of training water to run downhill, known in this tailor-made country as irrigating. You will see them up to their knees in mud and struggling like a toil trained cross between a beaver and a galley slave, and all of this toil worse than wasted.[26]

"Hard and discouraging" as it became, settlers finally learned better irrigation and farming techniques and meanwhile agonizingly wrested more acres from the sage and rocks each year—although never at the pace demanded by boosters who predicted that five million South Idaho acres soon would be irrigated. Between 1900 and 1920, irrigated acreage rose from about 609,000 to 2,488,806 acres.[27] Farmers planted orchards and grain and alfalfa and, surrounding their houses, rows of fast-growing Lombardy poplars which provided protection from winds that lashed the Snake River Plain in all seasons. Except on the Twin Falls South Side tract, described by William Jennings Bryan in 1907 as "desert land converted into a garden," and in the Boise Valley, where a variety of products came from farms, increasingly the cleared patches of land became largely alfalfa fields, supplying winter fodder for thousands of sheep and cattle that "Old West" barons grazed each summer on the mountain periphery of the Snake River Plain. "Now Will You Listen to the Gospel of Alfalfa," one journalist demanded of the farmers in 1909, to which he added, "Alfalfa is the groundwork of prosperity in all irrigated districts."[28]

These small beginnings—orchards, shade trees, growing crops, and stacks of hay—immediately were imperiled by uncounted hordes of cottontail rabbits (*lepus floridanus*) and their larger, more voracious jackrabbit cousins (*lepus compestris*). In one area, the *Buhl Herald* observed, the rabbits were "so thick that they are said to travel in flocks like blackbirds."[29] Until the first crops sprouted, settlers had mostly ignored the hares except as a source of food. But now the rabbits, particularly jacks, threatened ruin. Their natural habitat was uncleared sage-covered lands, but settlers complained that "no sooner does the sun disappear in the west than the jackrabbits begin to show themselves" to eat shade and fruit saplings and nibble on growing crops every evening. Jack molestations proved worse during the winter, when they devoured haystacks and consumed bark from trees. During the winter of 1909–1910, an abnormally cold season accompanied by heavy snowfall, jacks "completely barked" 1,500 apple trees of a single Acequia orchardist, and his neighbors suffered similar despoliation.[30]

"A Rabbit Drive. Fun for All But the Rabbits." *Twin Falls Public Library, Clarence E. Bisbee Collection, 667. Enhanced image courtesy Blip Printers, Twin Falls, ID.*

"30 Minutes Hunt," hunters display the results of a rabbit drive. *Twin Falls Public Library, Clarence E. Bisbee Collection, 2008.*

Settlers fought back in a twenty-year war with the jacks. They attempted to fence out the rabbits, buying expensive woven metal wire that many could ill afford. A Twin Falls-Salmon River tract settler recounted: "The rabbits nearly destroyed my crop in 1912. I had to fence against them and what crop was left lacked over $100 of paying for the fence. I had to go into debt for my living and feed for my workhorses for the next year."[31] But fencing proved no solution because rabbits burrowed beneath the barriers to reach crops and, in winter, easily reached haystacks when snow drifted against fences. To protect their trees, meanwhile, farmers experimented with coating tree trunks with offensive substances to repel rabbits. United States Department of Agriculture chemists recommended a mixture of lime, sulphur, carbolic acid, and "rancid soap suds." Doubtless angry farmers were tempted to make the coating even more unpalatable by adding cayenne pepper.[32]

The war far from won, settlers organized rabbit drives. Companies of people, arrayed in a giant semicircle, drove rabbits from the fields to the rim of deep canyons along the Snake and tributary rivers from which the animals plunged to their deaths. Elsewhere farmers erected V-shaped enclosures of woven steel wire with one open side, into which rabbits were driven before being clubbed and shot. By these methods, large numbers were slain, although writers undoubtedly exaggerated the slaughter—as when a 1906 drive near Hansen reportedly bagged 12,000 rabbits and 8,500 assertedly were killed in a 1914 drive near Acequia. More modest claims placed the kills at 1,000 to 4,000 per drive.[33]

At first, mainly settlers and their families participated in drives, but increasingly farmers tried to enlist aid from townsmen. However, cooperation from that quarter generally proved less than the farmers desired. To lure the townspeople, settlers then organized contests between teams of rifle-bearing sportsmen to compete for honors such as setting new records of numbers of jacks felled in one shot.[34] Encouraging such endeavors, one sympathizer contributed a bit of doggerel:

Bye, Baby Bunting
Daddy's gone a-hunting
Not for giants or for bears
But for deadly Bingham hares.[35]

Settlers also waged their fight in other ways. They condemned the Idaho Board of Land Commissioners, an agency whose policies scarcely ever pleased reclamation-tract residents. In this instance, settlers flayed the commissioners for leaving state-owned lands unoccupied and sage-covered; thus, the peppery *Heyburn Review* charged, "harbors of refuge" had been provided from which jacks continually invaded adjacent farms.[36] The commissioners responded, although never promptly enough to suit the complainers. By the end of 1914, the Board had leased 89,324 acres for agricultural uses, and it periodically sold other lands to farmers. In 1913 and 1914, for example, ownership of 35,505 acres of school and "special grant" land was transferred to private hands.[37] Portions of these lands were situated within the reclamation tracts, and the leases and sales helped abate the rabbit "refuges" that settlers denounced.

At the same time, settlers pressured boards of county commissioners to escalate the offensive against the jacks by paying bounties for slain hares. Several boards complied, offering bounties varying from fifteen cents for each rabbit scalp and attached pair of ears delivered to Canyon County authorities to three cents per scalp and attached ears presented to the auditor of Lincoln County. In the latter jurisdiction, commissioners reluctantly acceded to settler demands in 1910. In so doing, they risked scowls from influential stockmen about whom a reclamation tract dweller once commented: "When settlers on the farmsteads lock horns with the Cattle Kings and Sheep Barons of Lincoln County, something other than the Rams and Bulls is certain to get the worst of the controversy."[38] Finally, the carving of Twin Falls County from Cassia County and later the creation of Minidoka and Jerome Counties from portions of Lincoln and Blaine Counties eliminated much of the friction in south central Idaho over tax-supported rabbit bounties. County division, especially in the instances of Jerome and Minidoka, tended to segregate large stock raisers and reclamation tract dwellers in separate counties.

Clubbing and shooting rabbits continued, but by 1914 the practices were often questioned. Some townspeople criticized such methods as inhumane, and others called the rabbit drives "wanton destruction of a harmless animal." These opponents charged that farmers had grossly overstated their losses to justify the rabbit drives and urged that groups such as the Society for the Prevention of Cruelty to Animals intervene and help resist the agrarians' onslaught. Farmers, in response, amply documented the devastation that rabbits had wrought. The controversy came to a head in January of 1914, when Roy Palmer, deputy state game warden, arrested nearly a score of Porterville farmers and charged them with illegally shooting rabbits. Palmer cited an Idaho Attorney General's opinion that, if arms were employed in rabbit drives, the participants must possess valid state-issued hunting licenses. The Bingham County prosecuting attorney disapproved the arrests while Palmer's superiors, deluged with agrarian outcries, hastily reversed him. Scarcely mollified, farmers and their supporters interpreted the incident as misuse of state licensing powers to abolish rabbit drives. Then they demanded that the state of Idaho offer bounties for dead rabbits rather than "hamper their destruction "and informed other Idahoans that "anyone who wants to can [again] attend any of the rabbit drives without fear of being molested."[39]

State legislators parried the demands for state payment of rabbit bounties until 1919. Then lawmakers placed jacks on the list of predatory animals, a description with which settlers heartily concurred. However, the legislators authorized state-supported bounties only for six predatory animals that attacked sheep and cattle.[40] The disappointed warriors against jacks charged to no avail that the great stockmen had once more exercised their oppressive clout.

Obscured by these uproars were growing doubts among many farmers and their allies that bounties and rabbit drives alone were sufficient to cope with the jacks. Poisoning, by feeding the rabbits hay or grain containing strychnine, seemed more appropriate strategy, a view confirmed in 1913 by United States Biological Survey experiments near Cotterel and by Department of Agriculture demonstrations at Aberdeen and Kimberly. There scientists developed effective poison formulas and showed that jacks could be eliminated at a cost to farmers of no more than a penny per acre. The scientists' methods, too, appeased most humane minded critics who had objected to rabbit drives. Avoiding scientific jargon, one journalist explained the new methods to the humanitarians: "Mr. Jack finds himself in serious internal troubles which does not bother him for long."[41] Widespread use of poison followed and achieved the results promised by federal officials. County agricultural agents taught farmers how to use the poison safely, businessmen beamed as they profited from marketing chemicals and bags of poisoned oats, and then county agents organized and orchestrated campaigns that all hoped were the final showdown with the jacks. In Power County, for instance, agricultural agents persuaded the county commissioners to purchase 350 ounces of strychnine, which were distributed to the farmers in 1915. Agents estimated that 50,000 rabbits died that year in Power County. Elsewhere similar results were achieved. According to one report, "as many as 800 jackrabbits were killed in a single night on one farm," and large numbers of destructive ground squirrels were also eliminated.[42]

Poisoning and renewed rabbit drives at last brought the jacks under control, although rabbits still troubled farmers greatly on the periphery of the reclamation tracts nearest the deserts. There, a newspaperman reported, jacks remained as "dangerous [as] Comanche Indians or grizzly bears," especially when fires in summer destroyed their desert feeding grounds and deep winter snowfalls covered the vegetation beneath the sage.[43]

While most farmers perceived sage and jacks as enemies to be obliterated, some questioned their neighbors' resolves to destroy what might become valuable resources. A few eyed the hardy sage on unirrigated land, mused about its possible horticultural uses, and then conducted their own experiments. They grafted fruit and berry plants to sage, seeking thereby to produce fruit without labor and irrigation costs. The experiments failed repeatedly, but William Feen, an Emmett farmer, finally succeeded in grafting gooseberry and currant shrubs to sagebrush roots. Encouraged when the berry plants survived, he pruned some of his tallest and strongest sage and grafted on "pear scions" that reportedly "grew beautifully" the following spring.[44] Others believed that sage had greater potential for paper-making and perhaps rubber-extracting. Persons so minded remembered that two decades earlier Thomas A. Edison had theorized that sage was not worthless; instead it merely awaited man's ingenuity to find new uses for it—as did Robert Laing of Boise, who in the 1890's boiled sage in a lime solution and produced wood pulp from which he made paper.[45] Meanwhile, other Idahoans saw sage more promising as a source of chemical compounds and perhaps sophisticated medicines. Such theories received considerable backing because many Idahoans listened to folk teachings and accepted the notions of western explorers who first propounded these ideas in the nineteenth century. For instance, John C. Fremont remarked in 1842 that sage "saturated" the air "with the odor of camphor and spirits of turpentine," and the aromatic sage appeared "favorable to the restoration of health, particularly in the case of consumption."[46] Since then, moreover, Indian medicine men had made potions from sage which they used convincingly to treat typhoid fever among both their own people and the white newcomers. Based on Indian practices, folk *mate-*

ria medica then credited sage-derived substances with remarkable curative properties. Particularly in vogue for its reputed medicinal powers and tonic qualities was sage tea, a substance usually made from white sage (*artemisia gnaphalodes*) or fringed or pasture sage (*artemisia frigida*).[47]

After 1905, farm skeptics listened more attentively to these dreamers and finally desperately wanted to believe that their sage had "value other than a shelter for jackrabbits." Farmers grew excited because reports reached them that their large and infinitely more vexing *artemisia tridentata*, popularly called black sage, had great commercial value alongside the smaller sage varieties credited with medicinal and tonic properties. They were especially encouraged by a 1905 report that chemists had analyzed sage, finding in it "a large percentage of tannic acid, several alkaloids and a considerable portion of quinine."[48] No entrepreneurs, however, stepped forward in the wake of chemistry's new miracles to acquire any of the settlers' plentiful sage. Disappointed farmers then vowed never again to heed "blue sky" reports and returned to their view of sage only as an impediment to their progress.

Four years later, however, settlers found it difficult not to listen anew to those who represented sage as a valuable resource. Sylvester Sparling, a chemistry professor at Northwestern University, developed a process for economically distilling several useful compounds from sagebrush. Chicago entrepreneurs then formed the Chemical Products Company, to which Sparling continued to supply scientific expertise. In 1909, the company announced plans to raise at least $100,000 and establish "somewhere in the Twin Falls country a [manufacturing] plant for the utilization of sage brush." These promoters indicated they would extract wood alcohol, acetic acid, acetone, wood oil, creosote oil, pitch, carbonate of calcium and potash, and charcoal from the sage. Rumors about the enterprise spread and, their expectations raised anew, settlers prepared to market their once despised sage, becoming more confident of doing so within another year. In 1910, Idaho publishers reported that the Chemical Products Company had improved Sparling's process and now intended to focus on extracting tar, wood alcohol, and acetic acid from sage. The resulting "distillate," it was announced, yielded "a profit on the sagebrush of $15 a cord," and charcoal was an important and merchantable by-product of the process. According to the Idaho publishers, 4,000 pounds of sage yielded 220 gallons of chemical compounds and 350 pounds of charcoal, making Far West sagebrush worth "hundreds of millions of dollars."[49]

But again Idaho farmers were soon disabused of their illusions that sage was readily marketable. The enterprise, earlier announced for the "Twin Falls country," was lured to Nevada. The alert State Publicity and Industrial Commission brought Sparling to Reno in 1910 to demonstrate his industrial process; also, the commission secured independent appraisals from University of Nevada Experiment Station scientists, from scientists at two Utah universities, and from the Miner-Lawrie Laboratory at Chicago. All reported favorably, endorsing the process on both scientific and economic grounds. Then Nevadans, eager to attract a new industry, promptly donated free of charge 275,000 acres of sage to the Chemical Products Company, and several landowners agreed also to pay that corporation money for clearing their land of sage. In Washoe County, for example, the Madeline Valley Land Company offered a "bonus" of one dollar per acre when the Chemical Products Company removed the sage from 100,000 of its acres.[50] Small consolation to Idahoans it was when Thomas A. Edison suggested that charcoal, a by-product of Nevada sage-distilling operations, then be converted to "fuel gas" for use in Idaho. The famous inventor urged that the gas be supplied for engines installed alongside the Snake River to pump irrigation water onto the "wide flat valley" above the stream.[51]

To the great chagrin of Snake Plain settlers, local entrepreneurs then turned to modest enterprises processing sage leaves from which they extracted volatile oils that emitted sage odors. The oils proved useful in making perfume, hair tonic, shampoo, and other grooming products. In 1907, Shoshone businessmen organized the first such enterprise, the Sagebrush Hair Tonic Company. Boise, Mountain Home, and Twin Falls businessmen vied with each other to lure the company to their towns, but it stayed at Shoshone. Within a year, company founders built a factory where sage leaves were boiled, producing a clear green oil from which they made and marketed a substance for curing baldness, a hair shampoo advertised as curbing dandruff, and a soap for readily removing grease and other stubborn residues. Two years after its inception, an admirer claimed, the company had consumed "tons" of sage. However, the company declined in 1909 and 1910.[52] More unfortunately, such industrial enterprises never proliferated.

When Idaho sage did not become an important marketable resource, settlers speculated that their plentitude of jackrabbits might have unsuspected value. Little came, however, from their strivings to turn profits from the jacks. Only occasionally in the 1910s were settlers able to find buyers for rabbit pelts, and the few purchasers demanded that the skins be obtained in January and February when rabbits sported more luxurious furs.[53] Potentially more profitable was marketing jacks for human food. Settlers acknowledged that this rabbit meat was scarcely a delicacy, but it had proven edible: in past hard times, jacks probably were the principal meat consumed by many Idaho settlers.[54] Consequently, promoters tried—with limited success—to develop a market in eastern cities for dressed jacks. In 1911, for instance, a carload of 5,000 dressed rabbits, packed in ice, was shipped to a Pittsburgh commission house that had agreed to market the meat in that city.[55]

———♦———

By clearing their land of sage and rocks and preventing depredations by destructive rabbits, settlers had taken the first essential steps in taming their small corner of the Snake River Plain. Ahead of them lay incessant conflicts with irrigation companies and federal reclamation authorities for water enough to keep habitable that part of the plain so laboriously wrested. In some instances, as happened in the 1920s on the Twin Falls-Oakley and Twin Falls-Salmon River tracts, many settlers lost their claims to water and helplessly watched their acres revert to desert: there simply was not sufficient water for everybody on those tracts. Once crops in greater profusion appeared, then reclamation-tract dwellers and dry farmers wrestled with marketing questions so complex that sometimes there seemed no answers, especially in the long farmers' depression of the 1920s. Moreover, settlers dealt with all of these matters during periods of social ferment and upheaval created by pre-World War Progressive reformers, bent on bettering mankind according to Progressive lights, and then the disrupting influences of World War I followed. Preoccupied with their economic struggles, inevitably some settlers became short tempered when reformers intruded and forced settlers to cope simultaneously with economic adversity, changing social mores, and ideologies destructive of values once assumed to be eternal. Small wonder that one rural editor received so much support when he declared in 1914:

> If we are going to stand for our women to wear slit skirts, tight form-fitting dresses and vulgar hobble skirts, and our younger ones to dance the boll weevil, Texas Tommy tango, bunny hug, the bear dance, the kangaroo kick, the calf canter, the buzzard lope and so on down the line, the men folks might as well have their saloons and the whole family can go to hell together.[56]

Eventually settlers adjusted nicely to so much social change. "Progress," a Boise Valley booster said in 1920, "has lifted her magic wand," and the region's "citizens have sensed their destiny."[57] Meanwhile, economic adversity drove away by 1930 far too many reclamation tract settlers, proving in part Frederick Newell's reclamation adage of the 1910s that the efforts of three families were required to develop truly successful irrigated farms. But for those who stayed, many olden values seemed neither banished nor so outmoded as starry-eyed Progressives and conservative doomsayers proclaimed from diametrically opposite vantage points. Indeed, the Protestant work ethic appeared to the successful settlers exceedingly alive and well, as were Social Darwinist assumptions that survival of the fittest lay beyond mitigation. If those settlers who had stuck it out ever doubted these verities, publicists and public authorities hastened to reassure them. A writer remarked in 1917 of the Minidoka reclamation project: "Practically all of the ne'er-do-wells have been eliminated and their places taken by men who are able to take advantage of their opportunities and improve their own conditions."[58] In 1924, Warren Swendsen, Idaho Commissioner of Reclamation, congratulated Twin Falls North Side residents, telling them that they were "none but the most thrifty, energetic and enduring" for having triumphed in a "natural process of elimination."[59] At the least, Swendsen extolled human qualities demonstrably essential to prevailing over the sage and jacks.

This article originally appeared in *Idaho Yesterdays* 35, no. 1 (Spring 1991). Reprinted with permission.

Notes

1. *Biennial Report of the State Engineer to the Governor of Idaho for the Years 1899–1900* (Boise: Capitol Printing Office [1900?]), 8.

2. *Idaho: A Guide in Word and Picture*, 2nd rev. ed. (New York: Oxford University Press. 1950), 44.

3. J. C. Fremont, *The Exploring Expedition to the Rocky Mountains in the Year 1842 and to Oregon and California in the Years 1843–44* (New York: D. Appleton and Co., 1848), 93.

4. Howard R. Cramer, "Thousand Springs and Salmon Falls," *Idaho Yesterdays* (Fall 1974) 18/3:20; Merrill D. Beal and Merle W. Wells, *History of Idaho* (New York: Lewis Historical Publishing Co., 1959), 2:289–90; M. D. Beal, *A History of Southeastern Idaho* (Caldwell, ID: Caxton Printers, 1942), 319.

5. Cited in G. C. Hobson, ed., *The Idaho Digest and Blue Book* (Caldwell, Idaho: Caxton Printers, 1935), 314.

6. *A Victorian Gentlewoman in the Far West: The Reminiscences of Mary Hallock Foote*, edited by Rodman W. Paul (San Marino, CA: The Huntington Library, 1972), 29; Annie Pike Greenwood, *We Sagebrush Folks* (New York: D. Appleton-Century Co., 1934), 14.

7. For examples, see: William E. Smythe, *The Conquest of Arid America* (Seattle: University of Washington Press, 1969), 185–95; Greenwood, *We Sagebrush Folks*, 6–7; *Boston Transcript*, December 21, 1908, cited in *North Side News* (Jerome, ID), January 7, 1909; *Chicago Tribune*, January 17, 1909; *Chicago Record-Herald*, January 10, August 1, 13, 1909; "A Few Words about the Literature of Irrigation," *National Land and Irrigation Journal* (October, 1911), 4:21; C. M. Keyes, "A Country Ready for Capital," *World's Work* (August, 1909), 18:11,924.

8. *Denver Post*, October 31, 1909 (ten-page supplement on South Idaho); E. G. Adams, "America's Greatest Irrigation Enterprise," *Pacific Monthly* (November 1904), 22:286; Edward Dakin, "A History of the Buhl Herald Published Buhl, Idaho, 1908–38" (unpublished ms., 1938, Day-NW Collection, University of Idaho Library, Moscow), 1.

9. *Twin Falls Weekly News*, October 15, 1905.

10. Oscar O. Winther, *The Great Northwest: A History* (New York: Alfred A. Knopf, 1947), 326; Earl Pomeroy, *The Pacific Coast: A History of California, Oregon, Washington, Idaho, Utah and Nevada* (New York: Alfred A. Knopf, 1965), 375; O. H. Barber to D. W. Davis, April 1, 1920, David W. Davis Papers, Governors' Files, Idaho State Archives, Boise.

11. Report of R. B. Coglon, n.d. (1914?), Annual Narrative and Statistical Reports from State Offices and County Agents, Idaho, 1913–1944, Microcopy T857, Roll 1, Records of the Federal Extension Service, Record Group 33, National Archives (hereafter cited FES Records); Randall R. Howard, "Following the Colonists: An Account of the Great Semi-Annual Movement of Homeseekers," *Pacific Monthly* (May, 1910), 23:532.

12. Lloyd E. Byrne, *Buhl As it Was* (Boise: Syms-York Co., 1976), 11.

13. *North Side News*, August 4, 1910.

14. *Twin Falls Weekly News*, June 12, 1908; *Rupert Pioneer-Record*, March 18, 25, April 1, 1909.

15. *Biennial Report of the State Engineer…1899–1900*, 6; James J. Stephenson Jr., *Irrigation in Idaho*, Bulletin 216 (Washington: U.S. Department of Agriculture, 1909), 13–14; *Shoshone Journal*, May 17, 190; Greenwood, *We Sagebrush Folk*, 25; Annie Theresa B. Throckmorton, *Across the Years: "My Diary of Memories"* (Caldwell, ID: Caxton Printers, 1973), 140. Greasewood (*sarcobatus vermiculatus*) thrived best in soils too saline for largest sage to grow (John J. Craighead, Frank C. Craighead Jr., and Ray J. Davis, *A Field Guide to Rocky Mountain Wildflowers from Northern Arizona and New Mexico to British Columbia* (Boston: Houghton Mifflin Co., 1963), 194–95.

16. "Bruneau Extension Project: Report on Lands," July 29, 1911, Twin Falls Land and Water Company Papers, Idaho State Historical Society, Boise (hereafter cited TFLWC Papers).

17. Stephenson, *Irrigation in Idaho*, 58; Cloah Sebastian Cowling, *The Land of Sunrise Mountains: Memoirs of Idaho* (New York: Vantage Press, 1956), 129; Adelaide Hawes, *The Valley of Tall Grass* (Caldwell, ID: Caxton Printers, 1950), 42; *North Side News*, May 26, 1910.

18. "The North Side Tract, Southern Idaho," *Homeseeker's Illustrated Monthly* (April, 1915), 1:4.

19. Secretary of Twin Falls Land and Water Co. to Mrs. A. M. Merritt, August 3, 1906, TFLWC Papers; also see Dow Dunning, "Homedale and the Gem District," *Homeseeker's Illustrated Monthly* (July–August 1915), 1:30. The National Irrigation Congress warned that a "little money" was necessary but settlers with $1,000 or more were "almost sure of success" (*Twin Falls Weekly News*, August 1, 1907).

20. "The Minidoka Project," *Northwest Magazine* (August 1906), 2:166; *Twin Falls Weekly News*, April 7, 1905; *Rupert Pioneer*, January 3, 1907; Greenwood, *We Sagebrush Folks*, 25, 69, 85; George Stewart, *The Sowing and the Reaping* (Caldwell, ID: Caxton Printers, 1970), 70–71; I. G. Gooding, "Watermaster's Annual Report," December 31, 1932, Twin Falls Canal Company Papers, Idaho State Historical Society.

21. Earl Wayland Bowman, "The Twin Falls North Side Project: An Idaho Carey Act Project that is Robust and Secure," *The Golden Trail* (October 1916), 2:3.

22. Thomas Walsh to Glenns Ferry Canal Company, April 6, 1912, King Hill Extension Irrigation Company Papers, Idaho State Historical Society.

23. Bernard Grimes, "By Boxcar to Idaho," *Idaho Yesterdays* (Fall 1970), 14/3:20.

24. *Rupert Pioneer-Record*, August 19, 1909.

25. *Twin Falls Weekly News*, April 7, 1905, December 7, 1911; *Rupert Pioneer*, November 15, 1906. Throckmorton, *Across the Years*, 140: Ruth B. Lyon, *Valley of Plenty* ([Boise]: Capital Lithograph and Printing Co., 1968), 33; Gerhard Riedesel, ed., *Arid Acres: A History of the Kimama-Minidoka Homesteaders, 1912 to 1932. By People Who Were There* (Pullman, WA: G. A. Riedesel, 1969), 2.

26. Reprinted in Jerome *North Side News*, June 23, 1910; also see Charles H. McQuown to Fred A. Voigt, June 7, 1909, TFLWC Papers, and Alvis C. Holmes, *Swedish Homesteaders in Idaho on the Minidoka Irrigation Project* (Twin Falls: Ace Printing, 1976), 28.

27. Helen Lowell and Lucile Peterson, *Our First Hundred Years: A Biography of the Lower Boise Valley, 1814–1914* (Caldwell, ID: Caxton Printers, 1976), 103; E. F. Rinehart, "Demonstration Work on the Minidoka Reclamation Project in 1916," n.d., FES Records; U.S. Department of Agriculture, *Agriculture Yearbook 1924* (Washington, DC: Government Printing Office, 1925), 1105; Byron Hunter and Samuel B. Nuckols, *An Economic Study of*

Irrigated Farming in Twin Falls County, Idaho, Bulletin 1421 (Washington, DC: U.S. Department of Agriculture, 1926).

28. *Idaho Statesman* (Boise), April 25, 1976; *Twin Falls Weekly News*, September 27, 1907, April 30, 1909. Bryan, renowned "Boy Orator of the Platte" and U.S. presidential candidate in 1896, 1900, and 1908, was entertained at Ira Perrine's Blue Lakes ranch, visited several reclamation projects, and predicted success for all, *Shoshone Journal*, September 13, 1907.

29. Dakin, "History of the Buhl Herald," 5.

30. Addison T. Smith to W. B. Heyburn, July 26, 1905, Addison T. Smith Papers, Idaho State Historical Society; A. W. K. Kjoness, "Annual Report–1916," n.d., "Monthly Review of County Agent Work for December, 1916," n.d., FES Records; Cowling, *Land of Sunrise Mountains*, 123–27, William A. Bailey, *Bill Bailey Came Home*, Austin and Alta Fife, eds. (Logan: Utah State University Press, 1973), 67; *Twin Falls Weekly News*, May 5, 1905; *Rupert Pioneer-Record*, December 30, 1909, February 24, 1910, March 5, 1914; *Southern Idaho Advocate* (Burley), February 13, 1914.

31. M. A. Parrott to State Land Board, November 22, 1915. Moses Alexander Papers, Governors' Files, Idaho State Archives. Some rationalized the expense of fencing, saying it was worthwhile because the fences contained hogs and sheep.

32. *North Side News*, June 20, 1957; *Rupert Pioneer-Record*, February 24, 1910; *Southern Idaho Advocate*, February 13, 1914; U.S. Department of Agriculture, *Yearbook 1907* (Washington: Government Printing Office, 1908), 340.

33. *A Folk History of Twin Falls County* (Twin Falls: Standard Printing Co., 1962), 105. Riedesel, *Arid Acres*, 8; Byrne, *Buhl As It Was*, 17; *Rupert Pioneer*, February 1, 1906; *Emmett Index*, November 2, 1905; *Rupert Pioneer-Record*, January 20, 1910, January 8, 15, 1914; *Oakley Eagle*, January 10, 1913; *Pocatello Tribune*, January 3, 1914.

34. For examples, see *North Side News*, December 30, 1909, January 6, 1910.

35. *Bingham County News* (Blackfoot), December 31, 1920.

36. *North Side News*, January 13, 1910.

37. *Twelfth Biennial Report of the State Land Department of the State of Idaho 1913–1914* (Boise, 1914), 11, 12.

38. Mary Gunnell Lewis, "History of Irrigation Development in Idaho" (unpublished master's thesis, University of Idaho, 1924), 19; *North Side News*, January 20, 1910; *Rupert Pioneer*, February 13, 1908. Idaho laws authorized county commissioners to collect a tax of no more than five mills from which to pay bounties; the statutes set the minimum payment at fifteen cents for each rabbit destroyed: *The Combined Statutes of Idaho* (Boise: Syms-York Co., 1919) 1:998; USDA, *Yearbook 1907*, 561.

39. *Aberdeen Times*, January 8, 1914; *Twin Falls Times*, January 6, 1914; *Rupert Pioneer-Record*, January 22. 1914.

40. H.B. 213 and H.B. 214, *Session Laws of Idaho* (1919), 230, 568. The legislation placed cottontails on the list of game animals thus restricting any more rabbit drives against them.

41. *Rupert Pioneer-Record*, January 16, February 20, December 4, 25, 1913; *Southern Idaho Advocate*, February 14, 1914. In 1911, rabbit-poisoning experiments were conducted near Gooding by M. W. Smith, an Idaho deputy dairy, food, and sanitary inspector (*Idaho Statesman*, December 20, 1911). Prior to 1913, the U.S. Department of Agriculture frowned on poisoning because of dangers to humans, livestock, and wildlife (USDA, *Yearbook 1907*, 338–39).

42. "Monthly Report of County Agent Work in Idaho, November 1916," n.d., "Report" of O. D. Center, n.d., FES Records; *Aberdeen Times*, January 22, 1914.

43. *Bingham County News*, March 1, 1921; *Idaho Statesman*, July 10, 1978; Guy Flenner, *Out Where the Sage Brush Grows* (Boise: The Author, 1923), no pag.; *Aberdeen Times*, February 25, 1926.

44. *Rupert Pioneer*, March 8, 1906.

45. *Nevada State Journal* (Reno), February 3, 1892; *Elko (NV) Daily Free Press*, March 31, 1900.

46. Fremont, *Exploring Expedition to the Rocky Mountains*, 32.

47. Virgil J. Vogel, *American Indian Medicine* (Norman: University of Oklahoma Press, 1970), 396; Mildred Fielder, *Plant Medicine and Folklore* (New York: Winchester Press, 1975), 98–99: Ruth Ashton Nelson, *Handbook of Rocky Mountain Plants* (Tucson: Dale Stuart King, Publisher, 1969), 307; Burton O. Longyear, *Trees and Shrubs in the Rocky Mountain Region* (New York: G. P. Putnam's Sons, 1927), 232–33; *Elko Daily Free Press*, March 31, 1900; Blanche H. Soth, "Scent O' Sage," *North West Magazine* (July 1920), 11:392. For Idaho pioneer accounts about sage potions, see Ressia Roberts Helm, "Crossing the Plains in a Covered Wagon—1882" in Ann Nilsson Driscoll, *They Came to a Ridge* (Moscow: News-Review Publishing Co., 1970), 79; Bonnie Thompson, *Folklore in the Bear Lake Valley* (Salt Lake City: Granite Publishing Co., 1972), 99; Hawes, *Valley of Tall Grass*, 26.

48. *Twin Falls Weekly News*, September 29, 1905.

49. *North Side News*, April 22, 1909; *Rupert Pioneer-Record*, January 13, 1910.

50. *Reno Evening Gazette*, February 2, 1910; *Biennial Report of the State Publicity and Industrial Commission, 1909–1910* (Carson City: State Printing Office, 1910), 8–10. University of Nevada scientists reported that $23.81 worth of materials could be extracted from each ton of sage. Subtracting modest processing costs, the "net profit" per ton of sage was $20.56.

51. *Biennial Report of the State Publicity and Industrial Commission*, 11.

52. *Shoshone Journal*, December 13, 1907, April 17, June 19, July 31, December 4, 1908, May 14, June 18, July 9, 1909; *North Side News*, April 22, 1909; Rafe Gibbs, *Beckoning the Bold: Story of the Dawning of Idaho* (Moscow: University Press of Idaho, 1976), 52.

53. *Pocatello Tribune*, January 3, 1914.

54. For examples of human consumption of Idaho jacks, see: *Emmett Index*, November 2, 1905; Lewis, "History of Irrigation Development in Idaho," 19; *Pocatello Tribune*, January 3, 1914. Jackrabbit meat, a Department of Agriculture publication noted, compared unfavorably to meat from cottontails and domesticated rabbits, the jackrabbit "flesh, except in young animals, being somewhat coarse and dry" (USDA, *Yearbook 1907*, 334).

55. *Burley Bulletin*, December 15, 1911.

56. *Rupert Pioneer-Record*, January 22, 1914.

57. James R. Stotts, "An Idaho Eden," *The Golden Trail* (January, 1920), 4:11.

58. George Wharton James, *Reclaiming the Arid West: The Story of the United States Reclamation Service* (New York: Dodd, Mead and Co., 1917), 151.

59. Warren G. Swendsen, "Report on [the] American Falls Reservoir District" (Twin Falls: n.p., 1924), 22.

PART II

Irrigation Politics and Policy

Minidoka Dam, showing spillway and powerhouse, 1910. *Idaho State Historical Society, 77-127-2c*

CHAPTER 3

"Duty of Water" in Idaho: A "New West" Irrigation Controversy, 1890–1920

In the 1890s the Census Bureau declared the farming frontier closed for want of land suited to conventional agriculture. At the same time railroad promoters, irrigators, and journalists christened arid stretches of the Trans-Mississippi region the "New West," where agricultural riches were yet to be extracted and where man could still create "gardens" from "lava dust." When promoters miscalculated the New West's water resources and settlers misapplied the water, leaving some land water-logged while other areas suffered from water shortages, agriculture and irrigation scientists and state and federal reclamation authorities searched for remedies. Among the possible solutions were schemes for lowering the high duty of water that settlers upheld as essential to their success. A concept first devised by engineers, "duty of water" denotes the relationship between quantities of water and the land it irrigates. Water duty is high when a large amount is required to irrigate an acre of land, and it declines proportionately when smaller quantities suffice. Scientists in the early decades of this century urged state governments to impose low duties of water to force irrigators to reduce consumption and to make available water serve more acres. Such proposals interested officials in arid states and federal reclamationists. But when low duty of water proposals were introduced in Idaho, they generated one of the New West's most notorious irrigation controversies.[1]

Usually New West advertisers steered prospective Idaho settlers to the Snake River basin, an awesome plateau bounded by snow-capped mountains on its north and south perimeters and sprawling east-to-west for over three hundred miles across the southern part of the state. In 1890, the area's few human inhabitants resided in the Boise Valley, a farming oasis on the western extremity of the plain, and on irrigated farms scattered throughout the counties bordering on Wyoming to the east. The Snake River, its westward course semi-circular, bisected the unoccupied lands in the center of the plateau, but the meandering stream neither relieved the monotony of the landscape nor altered the hostile face of this dry, windy country whose first white explorers judged it an ugly desert. Sagebrush everywhere added to the bleakness, while on the western half of the plateau, the Snake River had carved deep canyons which impeded travel and rendered the terrain still more inhospitable.

Despite the harshness of the environment, few reclamation tract investors nor the thousands of settlers who reached the Snake River Plain between 1890 and 1910 questioned the region's water sufficiency or sensed the brewing crisis of the 1910s. An Idaho newspaper editor warned newcomers to "get down to solid, substantial facts" and ignore the "siren song" of boomers who would lure "many a poor family to disaster." But the new arrivals ignored his advice. Instead they listened to federal reclamationists who promised a rosy future on over four hundred thousand acres comprising the Payette-Boise and Minidoka reclamation tracts, and to local civic boosters who envisioned a "horizon full of rainbows." Idaho authorities offered settlers prospective Edens on over three million acres of supposedly reclaimable land which the state acquired under the federal Carey Act of 1894 and under two supplemental allocations. The land sold for fifty cents per acre, and by 1910 irrigation companies had laid out fifty separate agricultural tracts on which they anticipated placing $68,479,015 worth of irrigation structures.[2]

Indulgent state officials and Idaho courts further generated confidence that enough water was available to make the Snake River Plain bloom agriculturally. Because prior to 1903, officials imposed no "hard and fast rules" to prevent settlers from claiming water, irrigators routinely invoked "squatter sovereignty" to establish rights to vast streams of water. As long as their claims preceded those of newer irrigators, settlers had on their side the Idaho Supreme Court, which upheld the doctrine of "first in time, first in right." By 1900 the natural flow of the Snake River had been preempted thirty-five times, and the flow of tributaries had been similarly overclaimed.[3]

Idaho legislators intervened belatedly in 1903. To halt the practice of endlessly overclaiming available water, lawmakers imposed complicated procedures that supposedly would compel new water claimants to secure expropriation permits and then to demonstrate sustained and beneficial water usage before water rights could be awarded. Some water users, preferring the old squatter ways, simply ignored the law, while others found loopholes in the new statute that enabled them to circumvent the state's costly system and still assert their claims.[4]

Irrigation companies serving Carey Act tracts similarly sidestepped the new legislation and sometimes became the worst malefactors. The Carey Act of 1894 allowed companies profits amounting to the difference between their construction outlays for reservoirs, canals, and laterals and the amount charged settlers for water deliveries. Heady corporate competition for water customers ensued. To attract clients, irrigation companies established claims to large amounts of water and then justified their actions by loudly asserting the necessity of the highest possible duty of water for Carey Act project lands. At the same time, the companies endeavored to minimize their financial investment in water rights. A favorite tactic consisted of filing numerous applications on streams, only to allow the filings to elapse by not paying for permits to expropriate the water. Irrigation companies were able to repeat the procedure again and again and still assert their water priority in "time" and "right' ahead of other irrigators.[5]

Despite the frenetic claiming of stream flows, few settlers doubted the unlimited availability of water. Federal reclamationists and Carey Act project developers promised farmers huge storage reservoirs from which they could draw enough impounded spring runoff water to keep their fields perpetually green. The federally controlled Minidoka Dam across the Snake River seemed to support such claims since it regularly filled Minidoka Project gravity canals. Water shortfalls in 1910 seemed to demonstrate only a need for greater reservoir capacity. Farther downstream, Milner Dam similarly provided plenty of water for the Twin Falls South Side tract, a 244,000-acre Carey Act project in Twin Falls and Cassia Counties. William Jennings Bryan endorsed the Twin Falls South Side Project in 1907, and New West boosters cited its success as proof of their theories. Equally encouraging to settlers was the fact that the Idaho courts invariably decided water disputes in the irrigators' favor. Between 1905 and 1914, judges issued numerous decrees strengthening the claims of water users and setting priorities among users should water shortages develop. In part, this litigation was the result of a 1905 drought. Other lawsuits were initiated by settlers who, although convinced their water rights were "unassailable," took no chances and sought court decrees to insure their rights.[6]

Frequently, water adjudications paid land entrymen unexpected dividends. Judges often permitted litigants to "stipulate into the record" soil conditions and other circumstances which justified the use of water in excess of the two acre-feet judged adequate by irrigation engineers. Occasionally, courts awarded claimants quantities of water grotesquely disproportionate to actual need. In such instances the successful parties commonly attempted to sell their excess water to new entrymen or to the recipients of low water priorities. To protect themselves in event of drought, some farmers purchased the extra water. Others simply filed new lawsuits, gambling that jurists would readjust earlier decrees and grant the plaintiffs larger shares of the disputed water. For one stream, the courts adjudicated water rights five times in nine years, leaving ownership of the water still contested in 1914. Less fortunate settlers along the stream had a chance until 1921 to improve their water rights. With the stakes so high, unsuccessful litigants could not afford to despair. As State Engineer D.W. Ross acknowledged, "irrigation and litigation" were virtually "synonymous" in Idaho, and entrymen defeated in one lawsuit cheerfully prepared for another round by "selling [son] John's colt" to finance the court costs and attorney fees.[7]

Non-farm Idahoans generally applauded the irrigators' freewheeling acquisition of water rights, believing that such freedom sustained Idaho's reclamation boom. A vocal minority, however, strongly disagreed. Irrigation experts and thoughtful conservationists predicted disasters ahead without better regulation of access to Idaho's water resources. Many old-line stock raisers similarly looked askance at the rapid expropriation of streams, and viewed the state's reclamation frenzy as a mixed blessing at best. Together cattlemen and conservationists denounced the practices which encouraged reclamation tract settlers to act as though Idaho was—as nationally advertised—a land with nearly unlimited water at irrigators' disposal. To halt excessive irrigation, they demanded duty of water legislation restricting usage to two acre-feet of water per acre in each irrigation season, an amount which allegedly had been proven adequate by early Boise Valley farmers. State lawmakers not only refused to limit water usage, but on February 25, 1899, enacted a statute affirming that "the consumer of water shall be the judge of the amount and duty of water required for the irrigation of his land."[8]

In 1903, the Idaho legislature conceded somewhat to conservationists' demands and imposed an upper limit on water usage near the proposed two acre-foot standard. The new law stipulated "That when water is used for irrigation, no such license [to it] or decree of the court shall be issued confirming the right to use more than one second foot for each fifty acres of land so irrigated." The legislators, however, left wide loophole for irrigators to exploit by allowing courts and state reclamation officers (the State Engineer and the Board of Land Commissioners, or Land Board) unlimited discretion to award settlers larger quantities of water upon finding "a greater amount is necessary." Conservationists demanded closure of the loophole in the new statute and continued to campaign for an acceptably low duty of water guideline, but they faced insurmountable opposition from settlers who pointed out that no sound scientific studies of Idaho land and irrigation had been conducted to justify a two acre-foot limit. On the side of irrigators stood businessmen, lawyers, and many others profiting from Idaho's reclamation boom.[9]

Settlers exploited their advantages ruthlessly. Land entrymen linked arms with Carey Act reclamation tract promoters to convince the Land Board to award quantities of water as high as forty acre-feet per acre. Eventually, the board established 3.7

acre-feet delivered to settlers during a 150-day irrigation season as the annual per acre duty of water on Carey Act tract lands. Idaho courts generally upheld Land Board awards, or conveniently sidestepped duty of water questions by refusing to hold any particular duty acceptable. Judge George H. Stewart, ruling in 1905 on Boise River water rights, held that duty of water depended upon "actual necessity and economical use." His final decree of January 18, 1906, fixed no standard duty of water for land irrigated by the Boise River. Federal authorities likewise delivered copious water flows to farmers on federal reclamation projects. On the Minidoka tract, for example, Acequia district agriculturalists who argued that their soil was exceptionally sandy received water as much as twenty times above the proposed two acre-foot duty of water guideline.[10]

Deeply rooted customs, together with irrigation company practices, ensured hostility toward a restrictive duty of water. Custom held it unpardonable to deny irrigators less than "continuous [water] flow" during the irrigation season. In the winter months settlers asserted rights to additional canal water for livestock and domestic use. When such customs were denounced as possibly illegal and certainly wasteful, irrigators invoked their rights and pleaded crop "ruination," devastated livestock, and human privation. Settlers justifiably resisted drilling expensive wells to reach deep Snake River Plain water tables and pointed out that their land was unsuited for the more economical method of rotating heads of irrigation water among farmers. On Carey Act tracts, meanwhile, developers competing ruthlessly for settlers appealed regularly, and with Land Board approval, to potential entrymen by offering them substantially more than two acre-feet of water per acre. Some companies simply supplied their customers with "all [the water] they wanted."[11]

Boosters advertising the New West nationally won legions of influential supporters for reclamation of the Far West. Voices from the federal bureaucracy, notably those of Richard Hinton and John Wesley Powell, contributed significantly to New West boosterism. Over one hundred million desert acres, they estimated, awaited the hands of reclaimers. Greater stimulus came from outside the government, first from William Smythe and his associates. An *Omaha Bee* reporter, Smythe launched the National Irrigation Congress in 1890, and for nearly a decade that organization and its mouthpiece, *Irrigation Age*, relentlessly publicized the New West. George Maxwell founded the National Irrigation Association in 1897 and eventually became the New West's leading publicist.[12]

New West publicists joined irrigators, railroad agents, and other journalists to convince Americans that irrigation could work magic on arid lands. Impressed by such claims, financiers and investors bought irrigation works bonds and thereby helped initiate Far West reclamation projects. President Theodore Roosevelt's conservation gospel, likewise, whetted the interest of middle-class Progressives who pressured Congress for federal funding of reclamation tracts. Some of Roosevelt's disciples even left their comfortable Eastern homes to become New West pioneers. Also among the New West's allies were prominent metropolitan dailies like the *Chicago Tribune*, leading agricultural scientists, and even contemporary novelists Harold Bell Wright and Maude Redford Warren.[13]

New West publicity attracted additional thousands of settlers to Idaho. Between 1900 and 1910 the population of the state doubled to 325,594. A large proportion of these new settlers were former townspeople, particularly ex-businessmen and professional persons, who continued the freewheeling pursuit of water rights. Untutored in irrigation matters, they equated high volume of irrigation water with abundant crop yields and insisted upon the largest quantities of water possible, even though they frequently were unable to cope with the amounts allotted.[14]

Water suppliers continued to placate their clients and made little effort to calculate or control the amounts of water supplied. Irrigators and suppliers alike ignored Idaho statutes requiring measurement of water deliveries. According to the State Engineer, scarcely one-half of Idaho's irrigation water was measured in 1915, partly because settlers fiercely resisted installation of the required weirs and orifices. Irrigators resented being assessed the cost of installing measuring devices which many contended were inaccurate. Others, fearing crop losses, threatened watermasters with lawsuits if water curtailments resulted in smaller crop yields.[15]

Settlers became adept at taking advantage of the few diversion and measuring devices in canals and laterals. Some irrigators appropriated water by illegally digging holes in canal embankments. Others, known as "water hogs," waited until water reached laterals before expropriating it. Only neighbors' stern demeanors and prowess in pitchfork-and-irrigation-shovel litigation deterred such acts. Less assertive settlers irrigated at night while carrying lighted lanterns. Because federal Minidoka Project officials and irrigation companies on Carey Act tracts usually left division of lateral waters to the farmers, abuses proliferated. T. William Parry, a Minidoka Project hydrographer, reported in 1911 that irrigators "at the upper end" of laterals habitually took the water they wanted, while those at opposite ends of laterals could expect from one-fourth of their share to "none at all."[16]

Popular as these reckless practices were, they could not persist. The largest and best-watered tracts, notably the Twin Falls South Side and Minidoka projects, were soon fully occupied and project authorities refused on financial grounds to build more reservoirs. Instead, water users were told that they must irrigate more economically, so that the newest entrymen might share equitably in the available reservoir water. Patches of waterlogged land and creeping alkali deposits presented even more persuasive arguments for immediate water conservation. Alarmed project officials, and even some settlers, complained of excessive irrigation and seepage from canals during the winter months.

It was generally agreed that the first step toward curbing water misuse lay in delivering water only in the summer irrigation season, even though that alternative would force settlers to drill costly wells for their domestic and livestock needs. Despite farmer protests, winter deliveries gradually stopped during the 1910s. These curtailments, however, could not by themselves prevent land waterlogging, conserve enough water to supply all users equitably, and at the same time justify extension of existing projects—notably the Minidoka Project—irrigated land for which settlers demanded water.[17]

To the consternation of irrigators, their pesky water-conserving adversaries pronounced doomsday at hand and declared indisputable the need for a low duty of water. More worrisome to water users, the restrictive duty of water viewpoint had won converts since the turn of the century. Federal reclamation officials at Boise and Rupert, whose clients included the state's most prolific water users, saw new merit in the proposal. Most agricultural scientists also concurred with irrigation engineers who again endorsed a two acre-foot duty of water as economically feasible, scientifically tenable, and socially necessary. For their own self-serving reasons, many Carey Act tract irrigation companies likewise threw their support behind the two acre-foot standard. Under the lower limit, they could save a great deal of money that would otherwise be spent fulfilling contractual obligations for large amounts of water. Some Carey Act project developers also saw in the lower standard a solution to the dilemma of having sold entrymen more water than the company could provide.

For the first time, settler rights to self-determination on irrigation matters were truly threatened. Farmers and their boomer allies in town reacted vehemently, denouncing any guidelines that compromised the 1899 law which made irrigators sole judges of the "amount and duty of water" appropriate for their land. To substantiate their claims that agricultural success in Idaho depended upon a high duty of water, settlers invoked anew old arguments about soil quality and other mitigating factors.

Farmers and boomers gained the upper hand in the dispute when low duty of water advocates failed to produce scientific studies to rebut irrigators' claims. The case for low duty of water rested primarily on conservationists' unsubstantiated perceptions of irrigation in Idaho and other states, their seat-of-the-pants intuition, and very little scientific information. Conceding the unconvincing nature of the conservationist position, Pittsburgh financier William Kuhn, who had investments in three Idaho Carey Act projects, pressured state officials for scientific duty of water investigations, offering to pay part of the cost himself. Other irrigation companies—not to mention Governor James Brady, the Land Board, and Idaho lawmakers—however, refused to share the financial burden of elaborate studies. Finally, low duty of water advocates retreated, their own folly a major embarrassment to them.[18]

Pending empirical proof of the benefits of low duty of water in Idaho, conservationists turned next to water rotation as a method of curbing excessive irrigator appetites while at the same time allowing farmers sufficient water to irrigate their crops. Irrigators in Utah and Colorado had already accepted water rotation, and even in Idaho one Boise Valley irrigation company had imposed, despite farmer lawsuits, a water rotation system that had saved water and had not harmed crops. Watermasters, irrigation engineers, and state and federal reclamation officers all endorsed rotation, claiming that it allowed quick application of large heads of water to crops and that it eliminated waste. Rotation, therefore, not only would ensure high crop yields with minimal personal inconvenience to irrigators, but it would also reduce water usage to the low duty of water guideline that conservationists demanded. Irrigators disagreed and denounced water rotation as an "innovation" and an "infringement of their rights."[19]

In the face of strong farmer resistance, reclamation project officers and state and federal officials debated ways of implementing water rotation. Some experts suggested simply declaring a low duty of water essential and then summarily imposing rotation. So high-handed a course of action, however, invited judicial rebukes. Also, aside from provoking settler revolts, arbitrarily fixing a two acre-foot limit might affect irrigators inequitably. Finally, not all soil scientists agreed on a fair and sufficient duty of water. Studies showed, for example, that Twin Falls-Salmon River tract lands required a minimum of anywhere from 1.5 to 2.26 acre-feet of water "applied to the [farmer's] ground." These findings, of course, ignored Salmon River tract settlers' contractual rights to 2.75 acre-feet per acre.[20]

Nevertheless, compulsory water rotation was imposed on some Carey Act tracts. The Idaho Irrigation Company haphazardly rotated deliveries on its Lincoln County tract in 1910, and during the following year the Twin Falls-Salmon River Company staggered its deliveries. More important, Twin Falls South Side Project leaders imposed a rotation system that was carefully planned and well executed, and which enjoyed the endorsements of local water-user leaders. Although irrigators challenged the Twin Falls South Side system, the Idaho Supreme Court eventually ruled against them, holding that mandatory water rotation was acceptable "if necessity and the economical use of water required it." Citing the Supreme Court ruling, South Side and Twin Falls-Salmon River

Project leaders again forced rotation on settlers in 1912, tempting other Carey Act project officials to follow suit.[21]

Federal Minidoka Project authorities, on the other hand, believed that notwithstanding court opinions, old irrigator preferences made enforcing "iron clad" rotation rules impossible. Certainly Minidoka settlers vindicated that view. As a result of past profligacies, Minidoka entrymen in 1914 faced accumulated bills of about $772,000 for drainage works to siphon off excess irrigation water. And yet farmers scoffed at irrigation expert Sir William Willcocks—modernizer of the Nile, Tigris, and Euphrates river irrigation systems and consultant to Interior Department secretary Franklin K. Lane—who preached economical water usage. In a 1914 Interior Department-sponsored referendum, Minidoka tract settlers voted down mandatory water rotation by a margin of nearly three-to-one. Finally, project officials encouraged farmers to try voluntary rotation plans. Even when irrigation district president E. L. Riggs advised settlers that "no amount of kicking" would "steer us clear of the proposition," Minidoka irrigators only slowly opted for rotation.[22]

While proponents of water rotation continued to disagree on whether to compel or to educate entrymen to become economical users, certain federal reclamation leaders injected new ideas into the dialogue. Still upholding the desirability of water rotation and a two acre-foot duty of water, they suggested new methods of effecting settler compliance. Reclamation Service director Frederick Newell publicly outlined one alternative at Boise and at Caldwell in October of 1912.

Newell proposed that reclamation supervisors install measuring devices to meter all irrigation water and to prevent farmers from evading water assessments by pleading inaccuracies in their billings. Settlers would receive amounts of water that crop scientists judged sufficient for plant maturation, and such water would be made available at fixed and reasonable prices. If irrigators subsequently demanded additional water, it could be supplied but at a surcharge imposed to discourage excessive irrigation. Under his plan, Newell envisioned farmers leveling their fields, placing corrugations between crop rows, and installing proper ditching in order to stretch inexpensive water allocations. Their farms thus equipped to handle large but periodic water flows, irrigators could no longer plead that technical impediments precluded rotation.[23]

While settlers grunted their disapproval, irrigation company officers and some project managers expressed keen interest in Newell's proposal. In particular, they were impressed by tests of Newell's ideas on parts of the federal Boise and Minidoka tracts where the most modern measuring devices had been installed and their accuracy confirmed. On the southern sectors of the Minidoka Project, where irrigators depended on pumped water and necessarily subsisted on restricted water supplies, farmers grudgingly tolerated rotation, paid for their water, and were assessed additionally if they required more. Federal reclaimers argued that these measures had eradicated many irrigation abuses so commonplace elsewhere in Idaho and had virtually reduced the duty of water on the two projects to two acre-feet. Some Boise Project farmers applied only 2.13 acre-feet of water per acre as compared to an average of 3.53 acre-feet elsewhere in the valley. Minidoka tract farmers, irrigating with pumped water, in three years nearly halved the average duty of water to 2.01 acre-feet in 1913.[24]

Because soil quality and composition varied, however, the tests on the Minidoka and Boise projects failed to prove that a two acre-foot duty of water was adequate to irrigate the entire Snake River Plain. Furthermore, the results of the experiments had limited application on Carey Act tracts, where contracts fixed water prices and usually allocated far larger water allotments than Newell allowed. Still, Newell's water conservation schemes appealed to Idaho Land Board members and other state officials who supervised Carey Act tracts, many of which were beset with unexpected troubles.

In 1914, Governor John Haines decried the "horrible mess" of Idaho irrigation and lamented that the crisis gave the state a "black eye" extending "practically around the world." On troubled Carey Act tracts, irrigators who depended upon pumped water staggered under escalating electrical power bills which compelled them to curtail irrigation. The most severe problems plagued several of the largest projects. Reservoirs on some tracts impounded less water than engineers had projected, and in other instances connecting canals and laterals had proven deficient. In either case, grave water shortages ensued, depriving settlers of their anticipated harvests. Collapse of the national irrigation bond market in 1911 made it virtually impossible for irrigation companies to raise new capital to rebuild defective irrigation systems. Under the circumstances, water conservation measures offered at least partial solutions to problems on Carey Act tracts. If settler-company water contracts could be revised, Newell's plan might yet provide a means of forcing water economies.[25]

Although Land Board members had fretted since 1908 over the depletion of Snake River Plain water resources, the crisis of the 1910s caught state officials napping. The board had ignored its own sense of alarm to approve new Carey Act projects in the years 1908–1910 and had modified its policies only to the extent of authorizing as little as 1.5 acre-feet of water per acre for new Carey Act tract irrigators. Suddenly confronted with the post-1910 debacle, the board—like other Idaho agencies—at first procrastinated, denying any blame for irrigation project troubles. In part, official inaction stemmed from disbelief that Idaho's ten-year binge of Carey Act reclamation could end so ingloriously. Also, the Land Board shuddered at the legal implications of settler charges that the board had been negligent in its supervision of Carey Act projects. When state officials finally responded to the crisis, they not only endorsed Newell's ideas but also solicited direct aid from federal agencies.

Pleas for assistance reached receptive ears in the departments of Interior and Agriculture, and funds and experts were quickly dispatched to Idaho. To some degree, prompt intervention served federal reclamation interests. Federal scientists and technocrats, more readily than elective state officials, could insist that Carey Act tract settlers subsist on far less water. Revival of failing Idaho Carey Act tracts, in turn, would establish a powerful precedent for compelling water conservation on federal tracts. Reclamation officials needed such ammunition. Many Minidoka Project clients still resisted economical irrigation practices, and the government anticipated similar battles with Boise Project dwellers upon completion of the massive Arrowrock Reservoir in 1915 or 1916.[26]

Together federal agents, the Idaho State Engineer, and the Land Board mapped a course to save endangered Carey Act projects. First of all, federal technicians commenced a federal-state financed inventory of southern Idaho water resources, beginning with the streams serving Carey Act tracts with the worst water shortages. For this purpose, the U.S. Geological Survey quietly placed gauging devices in water courses to collect sufficient data to allow the bureau to project each stream's long-range irrigation potential. The Reclamation Service simultaneously initiated its own studies of Boise River feeder creeks and began an inventory of the northeastern watershed of the Snake River. Both agencies also carefully reviewed water rotation methods and other water-saving ideas which offered genuine prospects for reducing irrigation water consumption.[27]

In the end, federal and state officials addressed the controversial duty of water issue. To refute irrigator claims that vast amounts of water were essential for farming, to establish indisputably the exact amounts of water on which irrigators could subsist satisfactorily, and to provide scientific underpinning for a new social policy that would deprive irrigators of excessive water rights, reclamation officials launched scientific studies to gather accurate statistical information. Engineers and non-farm water conservationists, always losers in past tussles with irrigators, enthusiastically applauded the government's new course, expecting "science" to vindicate their two acre-foot duty of water proposal. Once the lower guideline had received its scientific imprimatur, they hoped to persuade state legislators to impose and uphold it, even over the avenging howls of irrigators. Irrigation companies that had earlier requested scientific investigation, and who now desired a showdown on duty of water, also supported federal and state investigators. Should the new studies contradict farmers' inflated views on water sufficiency and belie early Land Board generosity in fixing duty of water, irrigation companies stood a chance of inducing the board to revise downward the companies' contractual water obligation to settlers.

The Department of Agriculture's Office of Irrigation Investigations took charge of Idaho water studies and in 1910 dispatched Donald Bark, an agency engineer, to direct the survey. For the next four years, Bark designed and supervised experiments which, he believed, demonstrated conclusively the "reasonable" limits of irrigation in the state. By the end of 1913 the study had cost the government $54,259, and Bark the following year requested an additional $11,000 to add finishing touches to the investigation. The Agriculture Department subscribed one-half of the final subsidy, with the remainder coming from the state-administered Carey Act Trust Fund and from irrigation companies.[28]

On 529 parcels of land scattered across the Snake River Plain, Bark irrigated 3,600 acres of crops with different quantities of water at varying intervals. This "major investigation" aside, he completed ancillary studies of irrigation on Boise Valley farms and of water seepage problems on the Twin Falls South Side tract. Aggressive and tireless, Bark did not wait for completion of his experiments. Using preliminary data, he supplied the Land Board with estimates of reclaimable acreage on Carey Act projects, and he worked with officials attempting to spread limited waters over water-starved tracts.[29]

Bark's experiments disturbed the equanimity of irrigators who had grown complacent after successfully beating back all efforts to impose a low duty of water. For some agrarians, Bark's studies constituted another foolish, pseudo-scientific fad financed with their tax dollars. Many, however, feared that Bark, an avowed water conservator, would devise impractical and more fiendish innovations than the hated water rotation methods that had already been foisted on some settlers. Federal reclaimers, state officials, and irrigation companies, they suspected, had prejudged the duty of water and had coached Bark to square his findings with an intent to slash drastically farmers' water supplies.

There is no evidence to suggest that Bark deliberately tailored his techniques to make the results of his experiments coincide with a predetermined duty of water. However, Bark belonged to that guild of technocrats who would save the New West from its follies by manipulating the duty of water in the interest of stretching New West water resources to the limit. Like his professional peers, Bark on philosophical and technical grounds remained ill-disposed toward irrigator viewpoints. Predictably, his scientific explanations and fervent evangelism offered little solace for Idaho irrigators.

In 1912, Bark invited farmers to visit his experimental plots at Gooding to learn ways to subsist on less water. The exhibits and lectures at Gooding confirmed farmer fears that Bark would reduce their water usage. Their concern increased later that year when the State Engineer published Bark's preliminary results and conclusions. Facing farm skeptics everywhere, Bark attempted repeatedly to explain his experimental results. With missionary zeal, he lectured agrarians that new "knowledge" meant that farming need no longer be a "lottery."[30]

Although Bark agreed with farmers that soil composition and quality, climate, and types of crops could be major variables in water usage, he supported no deviation from strict water conservation and he reproved Idaho farmers for their liberal use of water. A proponent of water rotation, Bark opposed settlers who argued that sensible farmers would use whatever amount of water was necessary to achieve the best crop yields. Instead, he argued that the government should determine an "economic Duty [of water]" on a project-by-project basis. Bark's formula for computing the duty, based on land and crop values and crop production expenses, in his view proved mathematically the desirability of less irrigation.

Although Bark stopped short of fixing a universal duty for the Snake River Plain, his teachings nevertheless galled irrigators, who noted that he essentially confirmed the judgment of those who had long advocated a two acre-foot duty of water. Initially holding 1.97 acre-feet yearly sufficient for each "average South Idaho" clay-loam acre, Bark eventually increased the "proper duty of water" to two acrefeet. To maintain that low standard, he conceded, farmers must practice "careful husbandry," accept water rotation, and devote only half of their total acreage to orchards, potatoes, sugar beets, and other crops which required substantially more than two acre-feet of water per acre. As such, Bark's findings supplied political ammunition to water conservationists laboring for a low duty of water, preferably one fixed by statute penalizing irrigator excesses.[31]

In the eyes of many farmers, Bark's ultimate insult lay in assigning half of their land to hay plantings and animal pasturage requiring little irrigation. Agrarians resented on principle this meddling with their crop choices. Because dairying and stock raising held little appeal for farmers on small reclamation tract holdings without access to summer grazing areas, Bark's proposal seemed especially insensitive. Settlers also noted bitterly that alfalfa hay production, heartily endorsed by Bark and once hailed as basic to prosperity on Idaho reclamation tracts, had produced mixed results at best. Hay surpluses were common in South Idaho, and except during severe winters large stock raisers purchased only part of the hay crop and paid rock-bottom prices. Quarantines imposed to prevent the spread of weevil infestation, meanwhile, prevented farmers from marketing outside the state their surplus hay.

Angry settlers, their vehemence surfacing publicly in 1914, served notice that Bark's ideas were totally unacceptable. According to one farmer, Bark was no more than a "tin horn government man" who had ignored agrarian outlooks and needs, and instead had acted as "some pumpkins" who "knows it all." Another settler predicted that Bark's duty of water guideline could become "a very vicious and dangerous menace." Irrigators widely disputed Bark's calculations. Some, hurling back Bark's own words, took refuge in olden irrigator subterfuges of holding up soil quality and other conditions to justify the highest possible duty of water. Fortunate Kings Hill Project irrigators cited a 1914 Reclamation Service survey certifying that each of their Elmore County acres required from 3 to 3.7 acre-feet of water annually.[32]

Riled agrarians, meanwhile, attacked those most vulnerable to their wrath, particularly Governor John Haines, an admirer of Bark's technocracy. Their votes in 1914 contributed to Haines's ouster from office and his replacement by Moses Alexander. A progressive-minded Wilsonian Democrat, Alexander presumably sympathized with settler underdogs who expected him to thwart the imposition of any arbitrary duty of water on the state.

Although the duty of water controversy persisted over the next four years, it gradually lost intensity. Repeated challenges to Bark's techniques and conclusions indicated that no standard duty of water could ever be amicably agreed upon in Idaho. Governor

Alexander, his aides, and the Land Board—an agency dominated by Alexander's political opponents—contributed to the ongoing struggle. Between 1915 and 1918, Alexander sided with irrigators in opposing any enforced duty of water. A supporter of water rotation, the governor nevertheless dismissed the bulk of Bark's findings as inconclusive, and discouraged water conservationists who continued to advocate a low water duty.[33]

Instead, Alexander and the Land Board agreed with new State Engineer J. H. Smith, who looked askance at his predecessor's efforts to promote a low duty of water in Idaho. Smith doubted that either a single duty or multiple "economic" duties, one for each project as Bark had suggested, was practicable. In his opinion, neither the courts nor scientists could devise a satisfactory duty because it was impossible to compile data so "accurate and complete" as to be undisputed. "The proper duty of water," Smith declared, "varies from season to season" and at "different times during any irrigation season."[34]

More important, Alexander and his subordinates frowned on duty of water guidelines because they recognized that such limits probably could not be enforced on Carey Act projects. Pointing to the large water entitlements embedded in contracts between settlers and irrigation companies, the governor predicted that jurists likely would protect irrigators' rights. Indeed, on June 29, 1915, Federal District Court Judge Frank Dietrich estopped an irrigation company from collecting money from settlers until the company had supplied all water stipulated in its contracts. The following year, the Idaho Supreme Court went a step farther and placed in the hands of irrigators a tool for compelling companies to abide by the letter of their contracts. The justices held that companies must deliver all water contracted for, even though settlers refused to make certain payments. The same court ruled on June 1, 1922, that an irrigation company must supply water purchasers the, "amount provided for in the contract" or "pay damages for its default," unless company failure was "excused." On rehearing the case, the court allowed only "extraordinary drought" as a "legitimate defense" against incurring default penalties.[35]

Anticipating such rulings, Alexander insisted that to relieve the water shortfalls on Carey Act tracts, which had precipitated this latest battle over duty of water, would require a drastically different course than that endorsed by low water duty proponents. Accordingly, he proposed major reductions in the total acreage of several projects in order to ensure that settlers on the remaining land would continue to receive the full amount of water for which they had contracted. In the case of other troubled projects, particularly the Twin Falls North Side tract where reclaimers initially had planned to irrigate 261,945 acres, Alexander plotted the eventual reclamation of entire tracts without depriving irrigators of a drop of their contractual water entitlements. To secure the necessary water, Alexander invoked the 1911 Warren Act, a federal law which permitted non-federal projects to tap excess water impounded in Reclamation Service reservoirs. David Davis, Alexander's Republican successor in 1919, similarly opposed a restrictive duty of water in Idaho. Davis looked hungrily for new reservoir water to revive the state's bustling pre-1910 reclamation era.[36]

When Idaho officials retreated after 1915 from a low duty of water standard, authorities administering the state's two federal reclamation projects followed suit. Although engineers, hydrographers, and other technocrats surrendered slowly, project supervisors no longer took duty of water measurements and guidelines seriously. The Reclamation Service also relaxed its policy, merely calling for more "rational use of water" on federal projects. To achieve this modest end, the service disseminated irrigation information to settlers, collected stiff surcharges from irrigators who used excessive amounts of water, and promoted water rotation methods.[37]

Boise and Minidoka project authorities especially encouraged water rotation, declaring it mandatory in 1918 and enforcing their own rotation rules in 1919. At the Minidoka Project federal officials also instituted a unique "zone" system of water deliveries, taking into account the tract's diverse soils which ranged from highly sandy to clay. Water was apportioned accordingly, with sandy zones receiving the largest amounts. Irrigators unlucky enough to farm on loam or clay, objected to the zoning, while other settlers protested water rotation as a second injustice heaped atop the unpopular zone system. Resistance became so stormy that Minidoka Project manager A. B. Brown resigned in 1919 to escape settler wrath. Resistance collapsed momentarily in the wake of Brown's departure.[38]

Excessive irrigation gradually decreased in Idaho, with federal projects leading the way. Under pressure from state agencies to practice rotation, Carey Act project farmers in the 1920s generally fell into line. In 1919, Minidoka Project water-users' consumption averaged 3.3 acre-feet for each of 104,259 acres. By 1923, most Boise Project farmers accepted an annual "standard water right" of five-eighths of an inch per acre. Nevertheless, the decline in water usage was not so precipitous as to assure that water conservation practiced in wet years would ensure enough holdover water in reservoirs to see projects through dry years. Nor had easing of irrigator excesses eliminated the need for expensive drainage works to prevent land waterlogging. Reporting on the Boise Project in 1920, the Reclamation Service noted that "Certain portions of the project have become water-logged and rising ground water is threatening additional acres."[39]

In resolving such problems in the 1920s, reclamation project officials, engineers, agricultural scientists, and irrigation companies ignored the old low duty of water panacea. The drive for an Idaho water duty law was dead, and duty of water lingered only as a quaint measurement of water usage. However, Idaho irrigators who had successfully avoided duty of water guidelines had merely postponed government regimentation. During the 1920s, a new vigilance against irrigator excesses took root among reclamation tract administrators. Yakima (Washington) Project manager R. K. Tiffany heralded the new adamancy when he warned in a 1917 speech that irrigators must acknowledge that "the right of a man to over-irrigate was no more sacred than his right to irrigate." In Tiffany's opinion, irrigator rights "must be abridged, if it is for the benefit of the whole community." Increasingly intricate water rotation systems eventually produced the result that Tiffany demanded, a curbing of freewheeling irrigation practices. For a time, Senator William Borah and other sympathizers helped irrigators to defend their old wasteful preferences, but inexorably rotation system supervisors continued the struggle against greedy irrigators. Because they wielded rule-making powers and enforced those rules unilaterally, the technocrats ultimately were in a position to win the war for water.[40]

Farmer resistance proved fruitless. At a 1919 "indignation meeting" at Rupert, settlers resorted to "intemperate language" and forced the resignation of an unpopular administrator. But the rules which had triggered their protest remained in force. Boise Project settlers were equally unsuccessful in 1920, when they fumed that federal reclamationists required irrigators to "toe the mark" and "conform to the rules of the [Interior] Department to a very minute degree." In short, Idaho irrigators were back to square one in 1920, their right to self-determination in irrigation matters still in jeopardy. Ahead of them lay crucial decisions. They could either fight the new regimentation to a standstill, perhaps using tactics tested in past duty of water controversies, or they could bargain with reclamation tract authorities for democratic irrigation rule-making and access to greater amounts of water.[41]

For nearly thirty years, Idaho irrigators had battled a loose coalition of enterprisers, technocrats, government officials, and civic do-gooders who preached water conservation and invoked the public weal to circumscribe farmer self-determination in irrigation matters. These struggles left Idaho irrigators increasingly skeptical of technocrats whose "science" had been harnessed on the side of conservationist social policy. Farmers, no longer impressed with turn-of-the-century Progressive faith in the goodness of government regulation, harbored deep hatred for meddling federal and state reclamation bureaus. And finally, settlers had tired of dealing with irrigation company enterprisers.

The long conflict over duty of water in Idaho had hardened the resolve of irrigators to preserve their individualism, protect their water rights, and prevent corporate enterprise and government officialdom from scuttling dreams of a bountifully irrigated "New Canaan" on the arid Snake River Plain. These principles would guide Idaho irrigators through the next two decades as they struggled to recover ground lost to technocrats, to reduce the power of reclamation tract authorities, and to procure enough new reservoir water to irrigate generously their Promised Land. Today southern

Idaho's intricate system of irrigation reservoirs stands as the most visible monument to the persistence of early Idaho irrigators and their sons, who waged the New West's battle for water.

This article was originally published in *Arizona and the West* 23 (Spring 1981). Reprinted with permission.

Notes

1. Henry F. Cope, "Making Gardens out of Lava Dust," World Today, X (June 1906), 621–28.

2. *Caldwell Tribune*, July 27, 1907; *Rupert Pioneer Record*, April 14, 1910. Heber Q. Hale to James H. Brady, April 23, 1910, James H. Brady Papers, Idaho State Archives [ISA], Boise. Unless otherwise noted, all newspapers cited in the notes were published in Idaho.

3. James J. Stephenson Jr., "Sixth Biennial Report of the State Engineer to the Governor of Idaho, 1905–1906," 6, typescript, Frank R. Gooding Papers, ISA. *Biennial Report of the State Engineer to the Governor of Idaho, 1899–1900* (Boise: Capital Printing Office, n.d.), 90; *James R. Nielsen v. Ralph Parker and Frank Parker* (1911), 19 Idaho Reports 727–34; *Emma C. Youngs v. Daniel Regan* (1911), 20 Idaho Reports 275–80.

4. B. W. Oppenheim & I. W. Hart (compilers), *The Compiled Statutes of Idaho* (3 vols., Boise: Syms-York Company, 1919), 2:1583–97; James Stephenson Jr., "Irrigation in Idaho," U.S. Department of Agriculture [USDA], Office of Experiment Stations *Bulletin 216* (Washington, DC, 1909), 54–56.

5. For an account of irrigation company tactics, see Archibald C. Milner to C. C. Baird, October 20, 1910, Twin Falls Land and Water Company Papers [TFLWCP], Idaho State [ISHS], Boise.

6. Merrill D. Beal & Merle W. Wells, *History of Idaho* (3 vols., New York: Lewis Historical Publishing Co., 1959), 2:161–62, 183; Paul L. Murphy, "Early Irrigation in the Boise Valley," *Pacific Northwest Quarterly* [PNQ] 44 (October 1953), 184. *Twin Falls News*, August 28, 1908; *Shoshone Journal*, December 6, 13, 1907.

7. F. W. Hanna, "Water Supply and Use for the Boise Project, Idaho," 44–46, typescript, Project Histories and Reports of Reclamation Bureau Projects [PHRRBP], Boise Project, 1906–1921, Microcopy 96, Roll 24, Records of the Bureau of Reclamation [RBR], Record Group [RG] 115, National Archives [NA]. *Ninth Biennial Report of the State Engineer to the Governor of Idaho*, 1911–1912 (N.p., 1912), 26–27. *Caldwell Tribune*, August 10, 1907. *Biennial Report of the State Engineer...1899–1900*, 91; Earl Wayland Bowman, "The Lower Boise Valley Country: A Skeleton Sketch of Canyon and Ada Counties, Idaho," *Homeseeker's Illustrated Monthly* 1 (August 1914), 5.

8. General Laws of the State of Idaho Passed at the Fifth Session of the State Legislature (Boise: George Lewis, Publisher, 1899), 383.

9. Oppenheim & Hart, *Compiled Statutes of Idaho*, 2:1594. *Emmett Index*, April 20, 1905.

10. *Ninth Biennial Report of the State Engineer*, 224–25; Murphy, "Early Irrigation in the Boise Valley," *PNQ* 44:183–84; Neil H. Carlton, "A History of the Development of the Boise Irrigation Project" (master's thesis, Brigham Young University, 1969), 69, 71. "Report

of Operation and Maintenance, 1914," 23, typescript, PHRRBP, Minidoka Project, Idaho, 1910–1919, M-96, Roll 102, RBR.

11. *Biennial Report of the State Engineer...1899–1900*, 84. "Report of Operation and Maintenance, 1913," 12–13, typescript, PHRRBP, Minidoka Project, M-96, Roll 102, RBR. Ray P. Teele, *The Economics of Land Reclamation in the United States* (Chicago: A. W. Shaw Co., 1927), 307–308. *Rupert Pioneer-Record*, April 4, 1912. F. A. Voigt to American Canal and Power Company, November 28, 1908; Charles H. McQuown to Voigt, June 7, 1909, TFL-WCP, ISHS.

12. Paul W. Gates, *History of Public Land Law Development* (Washington, DC, 1968), 644–46; William E. Smythe, *The Conquest of Arid America* (Reprint ed.), (Seattle: University of Washington Press, 1969), passim; Martin E. Carlson, "William E. Smythe: Irrigation Crusader," *Journal of the West* VII (January 1968), 41–47; George Wharton James, *Reclaiming the Arid West: The Story of the United States Reclamation Service* (New York: Dodd, Mead and Company, 1917), xvi–xviii; Samuel P. Hays, *Conservation and the Gospel of Efficiency: The Progressive Conservation Movement, 1890–1920* (Boston: Harvard University Press, 1959), 9–10.

13. "A Few Words about the Literature of Irrigation," *National Land and Irrigation Journal* IV (October 1911), 21. For examples of publicity extolling Idaho's superior water resources, see Smythe, *Conquest of Arid America*, 186–87; Cope, "Making Gardens out of Lava Dust," *World Today* 10:624; "The Land of Opportunity," *Denver Post*, supplement, October 3, 1909; James Brady, *Idaho: The Land of Promise and Fulfillment* (Boise: Statesman Shop, 1910); *The Land of Splendid Opportunities: Idaho* ([Pittsburgh, PA]: Twin Falls North Side Land and Water Co., n.d.); *Idaho, The Land of Opportunity* ([Twin Falls, ID]: Twin Falls Land and Water Co., 1906).

14. *Jerome North Side News*, June 23, 1910. A 1912 census of 1,433 Minidoka Project residents showed that only 808 had been farmers or ranchers before coming to the tract. "Report of Operation and Maintenance, 1912," 68, typescript, PHRRBP, Minidoka Project, M-96, Roll 102, RBR. See also "Report of R. B. Coglon," Annual Narrative and Statistical Reports from State Offices and County Agents, Idaho, 1913–1944, Microcopy T-857, Roll I, Records of the Federal Extension Service, RG 33, NA.

15. *Ninth Biennial Report of the State Engineer*, 271. "Minidoka Project History, 1914," 324–25, typescript, PHRRBP, Minidoka Project, M-96, Roll 98; "Report of Boise Conference of Operating Engineers for Irrigation Canal Systems Located in Idaho, Oregon and Washington" (January 28–30, 1913), 86, mimeographed, PHRRBP, Boise Project, ibid., Roll 23, RBR.

16. *American Fall Press*, August 17, 1907. E. C. Kierstad to J. D. Hibbard, September 19, 1911, Kings Hill Extension Irrigation Company Papers, ISHS. Annie Pike Greenwood, "Letters from a Sage-Brush Farm," *Atlantic Monthly* 124 (September 1919), 314; Bessie Woodward reminiscence in Olive Groefsema, *Elmore County: Its Historical Gleanings* (Caldwell, ID: Caxton Printers, 1949), 390. Voigt to F. H. Buhl, February 5, 1909, TFLWCP, ISHS. "History of the Minidoka Projects, Idaho, to 1912" (2 vols.), 1:75–77, 136–137; and "Annual Report, 1911," 103, typescripts, PHRRBP, Minidoka Project, M-96, Roll 96, RBR.

17. For accounts of Minidoka settlers, see "History of the Minidoka Project, Idaho to 1912" 1:43, PHRRBP, Minidoka Project, M-96, Roll 96, RBR; and William D. Gertsch, "The Upper Snake River Project: A Historical Study of Reclamation and Regional Development, 1890–1930" (PhD dissertation, University of Washington, 1974), 144–45.

18. W. S. Kuhn to Brady, January 12, 27, 1909; Brady to Kuhn, January 21, 1909, Brady Papers.

19. *Shoshone Journal*, September 10, 1909. Mary G. Lewis, "History of Irrigation Development in Idaho" (master's thesis, University of Idaho, Moscow, 1924), 42–43. "Report of Boise Conference of Operating Engineers for Irrigation Canal Systems Located in Idaho, Oregon, and Washington" (November 21–23, 1911), 4–5, 31–34, mimeographed, PHR-RBP, Boise Project, M-96, Roll 23, RBR. *Ninth Biennial Report of the State Engineer*, 170, 194–95. *Burley Bulletin*, December 1, 1911. Alexander McPherson, "The Duty of Water—Its Economic Handling," *Better Fruit* 4 (November 1909), 40. F.E. Weymouth et al., "Report of the Drainage Board" [October 17, 1913], N.p., typescript, PHRRBP, Minidoka Project, M-96, Roll 97, RBR.

20. F. P. King to Idaho Board of Land Commissioners [IBLC], July 6, 1914, John M. Haines Papers, ISA.

21. Idaho Irrigation Company to IBLC, n.d., Brady Papers, ibid. *Buhl Herald*, March 23, 1911. Will H. Gibson to IBLC, July 10, 1911, James H. Hawley Papers, ISA. *Idaho v. Twin Falls Canal Company* (1911), 21 Idaho Reports 410–61.

22. "Minidoka Project History, 1914," 286, 329, typescript, PHRRBP, Minidoka Project, M-96, Roll 98; "Report of Operation and Maintenance, 1912," 14–15; "Report of Operation and Maintenance, 1913," 12–13, typescripts, ibid., Roll 102, RBR. "Drainage Notes: Idaho, Minidoka," *Reclamation Record* 5 (February 1914), 50–51. *Rupert Pioneer-Record*, September 18, 1913.

23. "Boise Conference" (October 24–25, 1912); "Opening Address of Hon. F. H. Newell" October 24, 1912); "Address of Hon. F. H. Newell" (October 25, 1912), typescripts, Newell Papers, Library of Congress [LC].

24. "Annual Report for 1914," 208, typescript, PHRRBP, Boise Project, M-96, Roll 8; D. C. Henney & F. E. Weymouth to Newell, October 24, 1913, PHRRBP, Minidoka Project, ibid., Roll 101; "Report of Operation and Maintenance, 1913," 23–24, typescript, ibid., Roll 102 RBR.

25. John Haines to E. E. Elliott, November 25, 1914, John Haines Collection, ISHS. For examples of conditions on Carey Act projects, see E. A. Wilcox, "Irrigation Pumping in Southern Idaho," *Electrical Review and Western Electrician* 62 (January 25, 1913), 181–82; Bruce L. Schmalz, "Headgates and Headaches: The Powell Tract," *Idaho Yesterdays* IX (Winter 1965–1966), 22–25. Will H. Gibson to IBLC, August 22, 1911; "Report of State Land Board Meetings" (November 8, 9, 11, 1911), Hawley Papers, ISA; "History: King Hill Project," N.p., typescript, PHRRBP, King Hill Project, Idaho, 1911–1926, M-96, Roll 58.

26. Federal planners hoped to limit water use on the Boise Project to no more than two and one-half acre-feet per acre, an amount Reclamation Service studies showed sufficient. "Payette-Boise Project, Idaho: Operation and Maintenance Report to January 1, 1914," 12, typescript PHRRBP, Boise Project, M-96, Roll 22; W. G. Steward, "Duty of Water Investigations, 1909–1910, 87, typescript, ibid. Roll 18, RBR.

27. Minutes of IBLC, April 21, June 3, 1910, Brady Papers; "[Land] Board Actions during Hawley's Administration," typescript, Hawley Papers; C. Clyde Baldwin to Stephen D. Taylor, March 10, 1914, Haines Papers, ISA. *First Biennial Report of the Department of Reclamation, State of Idaho, 1919–1920* (Boise: N.p., 1921), 22.

28. Don H. Bark, "Experiments on the Economical Use of Irrigation Water in Idaho," USDA *Bulletin 339* (Washington, DC, 1916), 1–2. Bark to IBLC, February 27, 1914, Haines Papers, ISA.

29. *Ninth Biennial Report of the State Engineer*, 213–15, 274–78. *Gooding Leader*, October

26, 1916; *Buhl Herald*, September 10, 1913. Bark, "Report of Seepage Investigations..." (August 12, 1912), N.p., typescript, TFLWCP; Bark to IBLC, September 23, 1912, Hawley Papers; W. G. Swendsen, "Engineer's Report on Blaine County Irrigation Company's Carey Act Project" (February 18, 1914), 49, typescript, Haines Papers, ISA.

30. Bark to James H. Hawley, July 5, 1912, Hawley Papers, ISA. *Ninth Biennial Report of the State Engineer*, 238–312. Bark, "Duty of Water in Idaho," in Report of Boise Conference of Operating Engineers for Irrigation Canal Systems in Idaho, Oregon, and Washington (January 28–30, 1913), 68–76, mimeographed, PHRRBP, Boise Project, M-96, Roll 23, RBR. *Buhl Herald*, May 28, 1914.

31. *Ninth Biennial Report of the State Engineer*, 235–312; Bark, "Experiments on the Economical Use of Irrigation Water in Idaho," USDA *Bulletin 339*, *passim*, especially 56–57.

32. *Buhl Herald*, September 18, 1913; December 31, 1914. "General Report on [the] Kings Hill Project and Kings Hill Extension Project, Idaho" [1914], 47–48, typescript, PHRRBP, King Hill Project, M-96, Roll 61, RBR. The name was subsequently changed to King Hill.

33. Report of the Idaho Irrigation and Drainage Code Commission to the Governor of Idaho, 1915 (Boise: Capital News Job Rooms, 1915), 27.

34. *Eleventh Biennial Report of the State Engineer to the Governor of Idaho, 1915–1916* (N.p., 1916), 44. When irrigation companies persisted in demanding duty of water of their projects, the Land Board ordered the State Engineer to comply. *Gooding Leader*, March 27, 1917.

35. Rome Adams et al. v. Twin Falls-Oakley Land and Water Company (1916), 29 Idaho Reports 357–76; W. T. Tapper and S. J. Hopkins v. Idaho Irrigation Company, Limited (1922), 36 Idaho Reports 78–107.

36. Moses Alexander to W. P. Rice, August 7, 1915; to IBLC, August 3, 1916; Minutes of IBLC, January 29–February 3, 1916, Moses Alexander Papers, ISA. Gertsch, "The Upper Snake River Project," 188–228. *Second Biennial Report of the Department of Reclamation, State of Idaho, 1921–1922* (Boise, 1922), 250–51.

37. "Excerpts from the Seventeenth Annual Report of [the] Reclamation Service" [1918], typescript, James R. Garfield Papers, LC.

38. "Minidoka Project History, 1918," 303; and "Minidoka Project History, 1919," 371–74, typescripts, PHRRBP, Minidoka Project, M-96, Roll 100, RBR. *Gooding Leader*, February 21, 1918. "Annual Project History, 1918," 142–44, typescript, PHRRBP, Boise Project, M-96, Roll 10, RBR. J. H. Lowell, "Rotation of Water," *Canyon County Farm Bureau News* (February 1918), 7. *Burley Bulletin*, May 16, 1919.

39. *Twenty-second Annual Report of the Bureau of Reclamation, 1922–1923* (Washington, DC, 1923), 66. B. E. Stoutemeyer to Miles Cannon, November 23, 1923, PHRRBP, King Hill Project, M-96, Roll 62, RBR. U.S. Reclamation Service, "Idaho: Boise Project and Related Features" (1920), 23, mimeographed, copy in Burton L. French Papers, Miami University Library, Oxford, Ohio.

40. "Report of [the] Sixth Annual Conference of Operating Engineers Held in Boise, Idaho, January 15–18, 1917," 41, mimeographed, PHRRBP, Boise Project, M-96, Roll 23, RBR. William Borah to J. A. Handy, June 15, 1918, William E. Borah Papers, LC.

41. *Rupert Pioneer-Record*, May 1, 1919. Oliver O. Haga to W. R. Wood, December 1, 1920, copy in David W. Davis Papers, ISA.

Cottonwood flume from upper side. Milner Dam & Main Canal: Twin Falls Canal Company, Snake River. Bisbee Photo, 1912. *Library of Congress Prints and Photographs Division, HAER ID,27-TWIF.V,1-177*

CHAPTER 4

The Carey Act in Idaho, 1895-1925: An Experiment in Free Enterprise Reclamation

The Carey Act of August 18, 1894, permitted states to acquire from the public domain up to one million acres of undeveloped arid land within their boundaries, provided that such land could be made agriculturally productive; each state would contract with the private sector for reclamation work and would supervise the projects. Corporations and individual entrepreneurs chosen for the work stood to profit from the sale of water to settlers on reclaimed tracts. Hence, in a sense, the Carey Act authorized a free enterprise alternative to federal reclamation and to the empires subsequently built in the West under authority of the Reclamation Act of 1902.

In the first thirty years after Congress enacted the Carey Act, nine western states formally accepted land grants, but few more than minimally advanced reclamation thereon. New Mexico, Arizona, and Nevada virtually ignored their entitlements, setting aside only 7,605; 13,745; and 36,809 acres for free enterprise reclamation. Colorado, Montana, Oregon, and Utah took the law slightly more seriously; there, acres dedicated to Carey Act reclamation ranged from 141,815 in Utah to 388,877 in Oregon. On the other hand, Wyoming and Idaho were unique: they attempted not only to reclaim their million-acre grants but also to acquire and reclaim supplemental allocations.[1]

Consequently, Idaho and Wyoming's experiments under the Carey Act merit especially careful scrutiny. Such examinations can show how free enterprise spurred technological changes in farming and, more important, can illuminate the role of nonfederal reclamation in the growth and advancement of western agriculture during the early decades of this century.

Enactment of the Carey Act was particularly timely for Idaho. It created an unexpected boom in the southern two-thirds of the state, which had stood still economically since 1890. Westbound pioneers generally bypassed the Snake River Plain that stretched for about three hundred miles from counties bordering on Wyoming to the Idaho-Oregon boundary. They disbelieved representations of the great agricultural potential of this Idaho "desert." Mining activities had declined precipitously throughout southern Idaho since the 1880s; forestry held little economic promise there; and only a few thousand residents engaged in subsistence farming, for the region lacked reservoirs that would insure dependable supplies of water for their crops. In pre-Carey Act times, therefore, the main natural assets of southern Idaho consisted of tillable land and untapped streams. The former went largely unused except for spring grazing and winter quartering of livestock; the latter contributed several million acre-feet of water each year to the Snake River and eventually to the Pacific Ocean.

At least in theory, the Carey Act would open the door to agricultural expansion by stimulating utilization of "wasted" water resources and untouched lands. It would procure quick conversion of a million acres of desert to farms, and it would do so without opening the public purse. New capital investments in high-priced irrigation structures and technologies promised progress, new businesses, and schools to serve growing towns and communities.

Nonetheless, Idaho's experiment with Carey Act reclamation began amid controversy. When Congress first passed this law, many believed that it merely authorized schemes to enrich reclamation project promoters. One skeptic claimed that Carey Act enterprisers sang a "siren song" to entice "many a poor family to disaster" on new irrigation projects. Opponents also warned that free enterprise reclamation threatened to alter old-time lifestyles and to disrupt the region's stock-raising economy. Cattlemen and sheepmen upheld this view. They feared that reclamation would force them to share streams that, for want of any competition, had always been exclusively at their disposal. And were a million acres actually reclaimed on the Snake River Plain, stockmen must inevitably surrender grazing privileges and livestock wintering space on large blocks of land.[2]

Stockmen's misgivings counted more than opposition from other quarters because cattlemen and sheep raisers had influence in the state legislature, but supporters of Carey Act reclamation replied with gospel-of-progress arguments, the logic of which neither cattle barons nor sheep kings could dispute without acknowledging their self-interest. Carey Act proponents in the state noted that, because the Snake River Plain sloped gently from east to west, southern Idaho was susceptible to relatively inexpensive conversions to gravity-irrigated farms. Businessmen like Ira Perrine countered resistance with other geographical facts to prove that the public interest was best served by reclaiming the Snake River Plain. In south-central Idaho, the river dropped into a deep canyon where it remained for the next two hundred miles; ideal reservoir sites lay upriver from the canyon. Water could be diverted from reservoirs at these places at reasonable costs to irrigate hundreds of thousands of acres situated on both sides of the canyon.[3]

Advocates of the Carey Act described numerous tracts of land suited to economical reclamation. The biggest plum consisted of 200,000 to 300,000 acres of flatland situated northwest of Idaho Falls, a site that came to be known as the Dubois project. These lands, despite their agricultural potential, had remained vacant because irrigating them required large-scale diversion of Snake River water. The size and cost of such an undertaking had always kept agriculture close to the river's course. It was argued that the Carey Act would change that.[4]

Sometimes denigrated as "silk hat" enterprisers or silk hatters, advocates of free enterprise reclamation—chiefly local business and professional men—built their case on several theories. First, they argued, agriculturists lost substantial quantities of water from the Big and Little Lost Rivers, which meandered into the "deserts" of central Idaho where they "wasted" into the aquifer that underlay the Snake River basin. Such losses were preventable. Similarly, torrents of water escaped yearly when melting snow in the Sawtooth Mountains filled the Big and Little Wood Rivers in spring and early summer. This water could be captured to irrigate the north-central quadrant of the Snake River Plain. Silk hatters likewise argued for irrigating farms by tapping Mud Lake, the Weiser and Portneuf Rivers at opposite sides of the state, and creeks that descended from mountain ranges separating Idaho from Nevada and Utah. Of the latter, even the Raft River interested Carey Act boomers, although water was never plentiful in the area and homesteaders and stock raisers legally monopolized it; enthusiastic reclamationists announced that "Boise and eastern capitalists" were "ready to take hold" of financing the enterprise as soon as state authorities authorized them to proceed.[5]

Silk hatters pointed out that, since southern Idaho's best natural resources consisted of land and water, regional advancement depended heavily on putting these resources to better use. Invoking the public weal helped to persuade state legislators: in 1895, despite stiff opposition in each chamber, majorities in both houses voted to accept the responsibilities and the million acres coming to Idaho under the Carey Act. U.S. Department of the Interior and state authorities worked out guidelines for supervising reclamation of the land. The rules vested power in the State Board of Land Commissioners, an elected body commonly called the Land Board, to contract with private entrepreneurs for placing irrigation systems on Carey Act projects and then to supervise projects from inception to completion. Once the board judged projects completed, it would release the entrepreneurs from their contractual obligations to the state and transfer control of projects to settlers residing on the tracts.[6]

The Land Board authorized its first project, the 6,752-acre Marysville tract in Fremont County, in 1898; and in 1899, the board accepted proposals from the American Falls Canal and Power Company to reclaim 57,242 acres in Power and Bingham Counties. But these first experiments disproved claims that free enterprise reclamation guaranteed quick success. On the Marysville tract alongside the Fall River, local landowners blocked all activity until 1904, after which date a succession of corporations took control of the project and went bankrupt in turn. It took nearly twenty years to bring 5,852 acres under irrigation. On the other hand, the American Falls Company project was judged completed in 1910. At this time, however, irrigation water cost settlers four times more than the price quoted at the outset. There was no reservoir to store water for the dry season, and in any case the company could not provide enough water to irrigate the entire tract. In short, financial insecurity, water shortages, and endemic conflict between enterprisers and farmers dominated the two projects. More fundamentally, such conditions publicly underscored the inexperience of private enterprisers and showed them to be especially wide of the mark in calculating how much water farmers needed and how much that would cost to supply.[7]

At the least, difficulties branded the Marysville and American Falls Company projects unsuccessful prototypes for large-scale reclamation. But many observers went a step farther, speculating that Carey Act projects were paper tigers and that only federal reclaimers were likely to transform the Snake River Plain into an agriculturally productive region. For their part, stock raisers relaxed because free enterprise reclamation seemed to flounder along with its threats to the open range.

Critics mistakenly assumed that successive setbacks at Marysville and dubious results at the American Falls Company project discredited free enterprise reclamation and probably heralded an end to the state's experiment. Instead, commercial clubs and other business organizations from Weiser to Idaho Falls joined silk hatters in demanding more Carey Act projects after 1902. By then it had become obvious that the federal government, under the Reclamation Act, intended to reclaim no more than 450,000 Idaho acres in the foreseeable future. Having designed a 120,000-acre project in Minidoka and Cassia Counties and set its sights on tapping the Boise and Payette Rivers to irrigate another 327,000 acres, the U.S. Reclamation Service was uninterested in other Idaho projects.

It resisted the well-orchestrated campaign for the 200,000-acre Dubois project, which had failed to acquire private funding. And it denied that other sectors of the Snake River Plain were as suited to recovery as the federal choices. For instance, USRS officials judged the Big Lost River region undesirable for reclamation purposes.[8]

By 1910, forces outside the state helped to build momentum for free enterprise reclamation and to legitimize Idaho's Carey Act experiment. Theodore Roosevelt's conservation gospel whetted appetites for carving farms out of more land than the government could bring under irrigation. Among leading newspapers in agreement with his views, the *Chicago Tribune* staged a "land show" to promote western reclamation. A *World To-Day* writer adverted to planting "gardens" where "lava dust" dominated Idaho landscapes. Friendly to agents like Harry Hollister, who represented Carey Act enterprisers, the publishers of the Chicago *Record-Herald* lauded free enterprise projects in Idaho. More important, William Smythe and other publicists characterized arid regions as the "New West" and focused national attention on areas awaiting transformation into rich irrigated farmland. In Idaho, Smythe claimed, water was the most conspicuous asset, and "supremacy" in the future lay with people who were wise enough to use it to convert deserts into farms.[9]

Fortunately, too, for exponents of free enterprise reclamation in Idaho, good times returned nationally after the severe depression of the 1890s, and prosperity made large-scale Carey Act reclamation more feasible than it had been. Eastern financiers and investors loosened their grip on their money and less critically appraised appeals for substantial investments of capital in the New West. Trowbridge and Niver, a bond brokerage house at Chicago, seriously considered underwriting the Marysville Canal and Improvement Company's bonds in 1905 and subsequently helped to capitalize the company's project in Idaho. Smaller fry bankers, businessmen, and professionals were also persuaded to enter the irrigation securities markets. A popular magazine informed investors that irrigation company bonds enjoyed "a very fair record." A commercial publication generally endorsed bonds that Carey Act irrigation companies issued; one of its analysts described such securities as no different from railroad stocks, utility company bonds, and other commercial instruments. Although a few were of "doubtful merit," investors need not "mistake the bad for the good, or accept the counterfeit for the genuine."[10]

Once attracting substantial funding appeared certain, Carey Act reclamation became irresistible in Idaho for two reasons. First, money could buy new irrigation technologies vastly superior to what the American Falls Company had initially used. Second, sufficient money would permit construction of costly dams and storage reservoirs, lack of which lessened the odds for success in the early projects. Irrigators on projects like the American Falls Company tract still filled their canals during periods of spring and early summer runoff from the mountains, then subsisted on less water for the remainder of the cropping season. Often in the warmest summer months, canal intakes stood awkwardly above the water trickling past, and this moisture was denied thirsty crops.

Responding to the upswing, silk hatters revised their old schemes for establishing Carey Act projects. First, they rescaled projects to include extra tens of thousands of acres; next, they petitioned the Land Board to "segregate" land, meaning to reserve it for Carey Act projects; and, finally, they cajoled eastern entrepreneurs and brokerage houses to build projects from their blueprints. They proposed, for example, to infuse life into the Dubois project after federal reclaimers bypassed it, to reclaim more than a million acres in the Twin Falls "country," and to transform into irrigated farms the better part of all flatland situated around the Lost and Wood Rivers.[11]

The first backers capable of attempting reclamation on a grand scale were Frank Buhl and Peter Kimberly, whom promoters had interested in a quarter-million acres called the Twin Falls South Side project. Buhl and Kimberly, steelmakers and fabricators of steel products in Pennsylvania, had just sold their holdings to J. P. Morgan and Elbert H. Gary's "steel trust." In 1902, they agreed to raise money for and superintend completion of the South Side project. Kimberly died soon after making the commitment, but Buhl persevered. He raised over three million dollars, built Milner Dam behind which was created a reservoir of Snake River water, constructed two major canals, and connected laterals to these canals for irrigating about 244,000 acres. The South Side project soon emerged as the most admired of all Carey Act projects in Idaho. A writer described it in 1908 as "one of the miracles of modern American life."[12]

Spillway gates and irrigation falls, southeast view, Milner Dam & Main Canal, Twin Falls Canal Company, on Snake River, 11 miles west of Burley, Idaho. *Library of Congress Prints and Photographs Division, 059567p*

Supporters of Carey Act reclamation cited the South Side project as proof that free enterprise could quickly place land under irrigation and that it could be trusted to provide the best in contemporary irrigation technologies. Indeed, Buhl enforced high standards on builders of the South Side irrigation system; they created solid and reliable works; and because the system was so efficient, water was plentiful enough in early days for Twin Falls Land and Water Company officials to boast that settlers could have "all they wanted." For its part, the Land Board decided that Buhl and his associates had fulfilled their obligations within five years. (Lawsuits from a faction of dissidents, however, delayed until 1910 the company's release from state contracts and the transfer of control of the project to settlers.)[13]

In 1910, the South Side project contained 1, 295 farms under cultivation, and since mid-decade, the population of Twin Falls County had soared from a handful to 13,543. According to its admirers, the project attracted predominately a "class" of people certain to keep it prosperous over the long haul. Incoming farmers, one observer noted in 1905, were "the very best," and all had money to pay for land, water rights, and homes. Entrymen paid the state fifty cents per acre for the land, plus twenty-five dollars per acre for water rights, payable to Buhl and Twin Falls Land and Water Company bondholders in return for their expenses in constructing the irrigation system. Emboldened South Side officials discouraged settlers who lacked "capital and that in cash" to survive until their farms were well established. Among the preferred clientele were disgruntled farmers (characterized by some as "knockers") who fled from the federal project at Minidoka and still possessed enough money to buy land and water rights elsewhere.[14]

Minidoka's knockers and others chose wisely in locating themselves on the South Side tract. Towns grew steadily there, and local commerce developed momentum. After 1910, however, some Twin Falls Land and Water Company bondholders complained that profits reached only $7 per South Side acre, a sum not "commensurate with the investment" of so much money in the project. Given its success, they thought they should earn bigger profits.[15]

Public resistance to Carey Act reclamation in Idaho virtually disappeared with completion of the South Side project. In earlier times, critics anticipated land frauds on Carey Act tracts, and they expected enterprisers to cut corners and place profits ahead of quality. Conditions on the American Falls tract seemed to substantiate their distrust of capitalism. But Buhl's Twin Falls Land and

Water Company operated honestly and fulfilled its obligations. Many associated with the enterprise profited from securing land while it was cheap and from operating new commercial establishments at Twin Falls and other towns. Onlookers never detected serious flaws in the South Side irrigation system, and they failed to argue convincingly that federal reclamation would have achieved such good results so quickly. Consequently, even those old-time forces that once opposed Carey Act reclamation mostly changed their minds. A stock raiser, Fred Gooding, spoke as eloquently as anyone when he pronounced reclamation of arid land an activity best suited to the "beneficent control of private capital."[16]

Partisans of Carey Act reclamation frequently compared the completed South Side project to federal projects in Idaho and concluded that free enterprise reclaimed land more cheaply and more expeditiously than government. Others shared this view. The Colorado state engineer John E. Field claimed in 1913 that the cost of private projects was "about one-half that of Government enterprises." Ample water and a dependable delivery system on the South Side project and relatively amicable relations between settlers and the Twin Falls Land and Water Company invited comparisons with the situation at the U.S. Reclamation Service's Minidoka project. There, an observer noted, settlers were "very bitter against the Government." According to a *Collier's* magazine writer in 1910, South Side farmers prospered while Minidoka farmers still waited for federal reclaimers to build an acceptable irrigation system and to provide other amenities. At Minidoka, settlers disputed with USRS officials about the price of water, devising and administering systems for operation and maintenance of sublaterals, and funding that delayed construction of important irrigation structures.[17]

Idaho's defenders of free enterprise reclamation were aggressively critical of the USRS. Besides lambasting the director, Frederick Newell, and his subordinates (a "Consolidated Aggregation of Hammer Throwers"), Carey Act partisans accused the Reclamation Service of machinations to monopolize all reclamation activities in the West, and some even suggested ways of dismantling the agency's domains. The Idaho state engineer A. E. Robinson proposed that states fight back unitedly against all federal efforts toward "centralization of government" and "the gradual taking away from the different western states their right of control of natural resources."[18]

Between 1905 and 1914, the Idaho Land Board received proposals for fifty projects involving irrigation systems totaling $68,479,015. Out-of-state entrepreneurs promised financing for many of these projects, and reclamation eventually began on half of them. Among the twenty-five authorized was the Payette River valley project underwritten by the Chicago bond brokers Trowbridge and Niver, who also promised financing to reclaim 83,649 acres in the Big Lost River region where their agents were building irrigation works in 1909. New York entrepreneurs organized the Idaho Irrigation Company in 1906, which was intended to reclaim about 168,000 acres in the Big and Little Wood Rivers sectors of central Idaho. However, the New Yorkers' timing was bad; they placed long-term Idaho Irrigation Company bonds on national markets too soon after the panic of 1907 subsided and found that investors refused to "separate themselves from their ready money" for any "investment that cannot bring returns for more than a year." But such setbacks were temporary. J. G. White and Company of New York City, an engineering and construction industry giant, took control of the Idaho Irrigation Company in 1908, sold the bonds, and undertook the project alongside the Wood Rivers. And smaller but more complicated in design, two Kings Hill projects of 27,120 acres altogether were begun in the bottom of the Snake River canyon in western Idaho in 1907–09. Boomers advertised these reclaimed lands as potential paradises for growers of fruit.[19]

Significant as were these developments, many silk hatters and out-of-state enterprisers saw the best opportunities for reclamation in the Twin Falls country. The administrations of the Idaho governors Frank Gooding (1905–1908), James Brady (1909–10), and James Hawley (1911–12) publicly urged free enterprise to build projects like the Twin Falls South Side project on another million acres in the area. Edwin T. Meredith, publisher of *Successful Farm-*

ing, a farm journal with offices at Des Moines, Iowa, responded to these calls. A director of the U.S. Chamber of Commerce and the Federal Reserve Bank at Chicago, later to be appointed secretary of agriculture, Meredith organized the West End Twin Falls Company, contracted with the Land Board to build irrigation works, received permission to sell water rights to 46,016 acres, and arranged with Trowbridge and Niver to market West End Company bonds. Although the bonds failed to sell, Meredith's project remained in business. State officials even winked indulgently when he violated his contract by selling water rights before he could deliver irrigation water.[20]

Also interested in Twin Falls country was the Idaho businessman Ira Perrine, who mapped out some 300,000 acres situated on the north side of the Snake River canyon opposite the South Side project. Trowbridge and Niver agreed to sell enough bonds to capitalize this undertaking, but when the Pittsburgh bankers James S. Kuhn and William S. Kuhn bid for it and paid his price, Perrine transferred the North Side project rights to them in 1906.[21]

At this point, the entrepreneurial audacity of the Kuhns was unsuspected in Idaho, but their verve never again went unnoticed. After buying out Perrine, the Kuhns next sought a larger tract, nearly half a million acres in the Twin Falls country. The state engineer James Stephenson Jr. concluded that they could do in three or four years what would take twenty "under ordinary circumstances." So the Land Board granted them approximately 435,000 acres to reclaim, and it listened when they asked for more. The Kuhns soon coveted land that the Idaho Irrigation Company expected to reclaim; they secured sites on the Snake River to generate hydroelectric power; and they competed with the J. G. White Company for first chance at reclamation and hydroelectric power rights in the Salmon River watershed. They not only contemplated invading regions north of the Sawtooth Mountains, but they also reviewed the silk hatters' far-fetched idea of running water from the Salmon River into tunnels cut through the Sawtooths and then onto the Snake River Plain. Neither the Land Board nor other state officials wavered in their expectations of the Kuhns, and the governor trusted them not to "take hold of any proposition" having "the least semblance of failure."[22]

Within two years the Kuhns had requested more land than the Land Board could grant. The Carey Act allocation barely permitted the board to accommodate its earlier petitioners for land. Consequently, Governor Gooding pressured Idaho's congressional delegation to seek another million acres from the government. With such an allocation, the land board intended to satisfy the Kuhns, to allow several smaller projects to begin, and to authorize the massive Dubois project in eastern Idaho as soon as silk hatters located financiers. Congress granted Idaho a second million acres on May 25, 1908.[23]

But even before it was granted, the supplemental allocation fell short of Idaho's needs because the land board had decided belatedly to authorize another large-scale project. This one, proposed in 1907 by Buhl, Perrine, and certain bondholders and officials of the Twin Falls Land and Water Company, was called the Twin Falls-Bruneau project—soon simply Big Bruneau—and was, in the end, the largest single Carey Act project ever attempted in the state. At first, Buhl, Perrine, and their associates requested 380,000 acres, but the Kuhn brothers counterproposed on March 14, 1908, and the competition for control of the area grew more heated. Both sides plied the Land Board with blueprints for ever larger-scale projects, and the Kuhns finally topped their competitors by proposing to reclaim 791,000 Big Bruneau acres. Two days after making Idaho's second land allocation, Congress granted Idaho another million acres, and the Land Board moved quickly. It chose Twin Falls Land and Water Company plans over those of the Kuhns and, in two separate actions, awarded the company 527,040 acres. The corporation planned to spend $22 million for Big Bruneau irrigation works, and it sought these funds from Wall Street bankers. Rumors circulated that J. P. Morgan advanced the money, but funds were not forthcoming, even after the corporation pared its request to $13 million. (Lack of capital eventually forced abandonment of the project before work ever began.)[24]

Idaho's boom in Carey Act reclamation crested in 1912–13, although the Land Board subsequently received many applications for less ambitious projects, such as the Wickahoney Land and Water Company proposal to irrigate 40,000 acres with Bruneau River water. By 1914, the board had authorized twenty-five projects. That year it judged four of them "dormant" for want of progress; another five advanced past the "proposal" stage in 1915. Only the American Falls and Twin Falls South Side projects were completed. In all, architects of these projects planned to irrigate twenty-nine tracts at a cost of $54,172,655, of which $24,236,153 had been spent by 1916. According to the Idaho land commissioner S. D. Taylor, about 333,000 acres were "under actual cultivation and irrigation" in 1914, and enough water was available to irrigate another 775,000 acres. In large part, uncultivated land within projects belonged to speculators who had established their rights by meeting minimal "prove up" requirements and "had not been seen since." Not farmers, they simply waited for land values to rise before selling at a profit.[25]

State officials proclaimed Carey Act reclamation a success on several grounds. First, significant amounts of land and water had been newly utilized in the last decade, and the goal of bettering southern Idaho's economy by putting land and water to work seemed fully achievable. Social consensus backed up such assessments. Second, even the most diehard skeptics could not convincingly gainsay the presence of new technologies on Carey Act projects or discount the role of technology in advancing agricultural productivity on the Snake River Plain. Third, new irrigation works on Carey Act projects were vastly superior to old-time systems. Water now flowed whenever needed from new reservoirs into wide and deep canals that crisscrossed projects, and for the most part, these systems functioned with mechanical regularity.

Defects in Carey Act irrigation works did occur. Structural weaknesses developed in dams at the Big Lost River and Idaho Irrigation Company projects. Seepage of water from the Salmon Falls Creek and Little Lost River dams and from reservoirs within the Twin Falls North Side project proved difficult to control. Also, on the Kings Hill projects in the Snake River canyon, a complex system of flumes for conveying water over uneven terrain, of canals necessarily placed on dry hillsides, and of syphons for boosting water over hills repeatedly malfunctioned.[26]

Such problems were generally attributed to grafters in state government. Allegedly, the Land Board and other state officials maintained loose ties to Carey Act entrepreneurs and looked the other way when they saved money by creating substandard irrigation facilities. Critics accused board members of administering Carey Act affairs with "gross incompetence," "indifference to the rights of settlers," "favoritism to promoters and to political henchmen," and an eye to "feathering their own nests." The gubernatorial campaigns of James Brady and James Hawley reiterated such allegations in 1910. Reformers unsuccessfully urged the legislature to transfer Carey Act affairs from the Land Board to an independent commission. Democrats claimed that Republican administrations had illegally placed Carey Act monies in the state's operating budget, and in 1914, the governor-elect, Moses Alexander, called Republicans to task for their stewardship over this money.[27]

Although accused of corruption, the Land Board was chiefly guilty of making poor technical judgments about irrigation. Analysts often blamed the successive state engineers, the board's main adviser, none of whom was an irrigation engineer. As early as 1910, the board had uncritically accepted calculations of enterprisers on quantities of water essential to farming. Especially after speculators sold their holdings to farmers, not enough water was available on the American Falls Company and Twin Falls South Side tracts; dams for the Wood and Lost Rivers projects never collected sufficient water to irrigate all the farms; and water shortages on the Kuhn brothers' three tracts became notorious. In the latter instances, dams across Goose and Salmon Falls Creeks impounded water enough for only a quarter of the acreage to be served, and the Twin Falls North Side tract, even after being scaled down, lacked sufficient irrigation water.[28]

Relatively simple remedies were at hand for the American Falls Company and South Side shortages. The farmers who controlled the projects after completion simply purchased water from the U.S. Reclamation Service. The agency released surplus water, impounded for the Minidoka project at Jackson Lake, Wyoming, into the Snake River (over protests from the Wyoming state engineer A. J. Parshall).[29]

On the other Carey Act projects, contracts between the state and enterprisers were still in force. In cases like the Kings Hill tracts, the only practical remedies consisted of reconstruction of faulty irrigation systems; elsewhere, resolving problems entailed raising the dam at Jackson Lake's outlet, building a dam at American Falls, constructing a reservoir on the south fork of the Snake, and damming up the Payette River at Black Canyon. Old and even less viable solutions—like transporting Salmon River water in tunnels cut through the Sawtooth Mountains to the Snake River Plain—resurfaced. Meanwhile, water shortfalls placed the Kuhns in a dilemma. Either they scaled down the affected projects and lost revenue in proportion to the acres deleted, or they built new canals heading at the Snake River and risked diminishing profits.

Finally, recapitalization of Idaho's Carey Act experiment became unavoidable. The Kuhns and several other enterprisers attempted to raise money by selling new issues of bonds from their Idaho irrigation corporations. But purchasers of securities were elusive, especially when they heard in 1912–13 about financiers of Colorado, Idaho, and Oregon reclamation projects going bankrupt. The first to fall, Trowbridge and Niver teetered in 1911 and collapsed in 1912 as a direct consequence of underwriting the Denver Reservoir and Irrigation Company. Farwell Trust Company of Chicago then followed suit. At the Kings Hill tracts, the operating companies became insolvent, Farwell Company brokers failed to reassure creditors, and the latter forced the projects onto the auction block. Because there were no other bidders, the state of Idaho took possession of the projects for a pittance, then expended about $100,000 on improvements. Meanwhile, in 1913, the Kuhns' industrial and financial empires toppled under assaults in Pennsylvania; cut adrift, bondholders in the Idaho irrigation companies became the heirs to the Kuhns' Carey Act contracts to reclaim 435,000 acres.[30]

These bankruptcies, minor events in the larger world of American finance capitalism, indicated that the chances of recapitalizing Idaho's Carey Act experiment had become negligible. An officer of Porter, Fishback and Company, investment bankers at Chicago, analyzed investor perspective in these times. Entrepreneurs, he wrote, had attempted "the unnecessarily heroic in irrigation" only to discourage buyers of irrigation securities by flooding markets with insecure bonds in their haste "to open up territory" for which development should wait "for a generation." Investment Bankers Association officials agreed with this view.[31]

Nonetheless, Idahoans attempted to find new financing for reclamation of the state's three-million-acre land grant. The Land Board declared bondholders of old Carey Act irrigation companies accountable financially for pre-1913 agreements of enterprisers such as the Kuhns. Eventually, Twin Falls North Side Land and Water Company bondholders paid $447,000 out of their pocketbooks to purchase 315,000 acre-feet of water storage space at Jackson Lake. On another front, the state legislature established the Irrigation Securities Commission in 1913, its mandate to dispel distrust of stock and bond issues of irrigation corporations. The commission appealed to bankers and brokers at Cleveland, Chicago, Boston, and New York to reconsider Idaho irrigation securities, and Idaho commercial organizations demanded that the Land Board augment the commission's lures by offering all possible inducements.[32]

State authorities also tried to recapitalize the Carey Act experiment in another way. They urged settlers on Carey Act projects to organize irrigation districts that were empowered by state law to issue bonds, tax real estate, and pledge such revenues to redemption of district securities. The legislature amended irrigation district laws in 1913 to make irrigation district securities more attractive to buyers. More legislative tinkering with Idaho law codes followed, and finally in 1921, lawmakers established the Reclamation District Bond Commission. This agency obtained independent appraisals of irrigation district bonds and certified the bonds to purchasers on the basis of such ratings; moreover, certification officially informed investors that the "faith and credit" of the irrigation district stood behind the bonds. For a brief period, this legislation helped to make irrigation district bonds more marketable, but investors developed new mistrust because districts took advantage of the law and piled up too much debt.[33]

Meanwhile, all bonding strategies were unworkable for a handful of Carey Act projects that state officials judged failures. Included in this group were Edwin Meredith's West End Twin Falls project and the two Kings Hill projects. Governor John Haines (1913–14) requested federal takeover of the West End project in 1914, but the secretary of the interior, Franklin Lane, declined. Haines next attempted to interest Lane in the Kings Hill projects, but without success.[34]

Republican Party forces dominated state government until 1917. Republicans asserted that most of the three million Carey Act acres would eventually be reclaimed. Their optimism rested on high wartime crop prices, which lured settlers onto irrigation projects throughout the West. Possibly this resurgence in markets for irrigated farmland would have encouraged investors to open their pocketbooks at last and recapitalize troubled Carey Act projects. But postwar agricultural depressions affected, and raised questions about, the viability of reclamation, whether federal, Carey Act, or other private ventures.

A Democratic island in a Republican sea in 1915–16, the new governor, Moses Alexander, dissented from Republican recovery ideas. Partly because studies of the Idaho Irrigation and Drainage Code Commission impressed him, he favored shortening sail, concluding the Carey Act experiment quickly, and settling for reclamation of fewer than three million acres. He proposed several reforms that pleased settlers on Carey Act tracts: he advocated better administering of the act, promised redress of certain grievances, and appeared to sympathize with settlers who claimed to be victims of "legalized robbers."[35]

Reelected in 1916 along with a Democratic majority in the legislature, Alexander clarified his ideas on Carey Act affairs. The Democrat-controlled Land Board agreed with him when he called for abandonment of the three-million-acre land grant except portions for which sufficient irrigation water was available. He insisted on deleting tens of thousands of acres from the Big

Lost and Wood Rivers tracts, the Twin Falls-Oakley tract, and the Twin Falls-Salmon Falls Creek project. He closed down all projects that had "only a speculative interest." Alexander won his biggest fight with Carey Act enterprisers when he forced cancellation of the Big Bruneau project and returned the land to the public domain. Taking up a matter his predecessor had left unresolved, he demanded that the Wilson administration stand behind Idaho Democrats, whose election campaigns had promised federal annexation of the Kings Hill projects, and "make our promises good." In his battle with the federal government, Alexander was like "a man trying to let go of a bear's tail," but he got what he wanted. The state of Idaho formally deeded the Kings Hill tracts to the United States in 1917.[36]

———•———

Alexander actually attempted to save as much as possible of Idaho's Carey Act land grant. Invoking the Warren Act of 1911—a law authorizing sale of federal water to nonfederal projects—he urged the USRS to impound more water at Jackson Lake and release the extra to downstream Carey Act projects. The Reclamation Service declined. Under such circumstances, therefore, Alexander then continued to scale down projects over protests from unlucky farmers, land speculators, and irrigation company bondholders. Inevitably, some groups lost land, water rights, or revenue from selling water. Salmon Falls Creek and Oakley project settlers filed multiple lawsuits, and groups in the Big Lost and Wood Rivers regions agitated for federal or state intercession to transfer water from the Salmon River watershed to these projects.[37]

In the end, Alexander stood his ground, and his successors in the 1920s usually upheld his policies of requiring project sizes to fit water resources at hand. Consequently, the Carey Act experiment wound down, and the state abandoned most of the three million acres. In 1919, moreover, supervision of Carey Act reclamation passed from the Land Board to a new department of reclamation answerable to the governor, and this "reform" enabled Alexander and his successors to bypass the Land Board and the silk hatters attempting to revive the Carey Act approach. An exception, Governor David Davis (1919–22) regained possession of Big Bruneau lands that Alexander had relinquished, and he helped promoters look for project financiers. Critics accused him of paving the road for "eastern capitalists" to "exploit" Big Bruneau settlers. However, silk hatters never found the capital, and Davis, after becoming U.S. commissioner of reclamation in 1923 and effecting noteworthy improvements in many Idaho reclamation projects, changed his mind about supporting the Big Bruneau scheme. He favored dealing first with the problems of other Carey Act tracts and overcoming hurdles to completing reclamation on the federal Payette-Boise project.[38]

When the Carey Act experiment concluded, twenty-three projects remained functional, with 850,000 acres under cultivation in 1924. Furthermore, Carey Act reclamation had expanded the economy of southern Idaho significantly. By one estimate, "crop returns" from the Twin Falls South Side project alone put $40 million into the economy from 1905 to 1919. The Land Board calculated that between 1898 and 1913, Carey Act reclamation had attracted $100 million from out of state and was responsible for a population increase of 50,000 people. Other state authorities estimated that enterprisers had spent $24.5 million on irrigation systems; settlers had also invested their own funds permanently in their Carey Act land.[39]

By way of comparison, federal reclaimers brought about 120,000 acres under irrigation at Minidoka during this thirty-year period, fully reclaimed 144,000 acres, and provided another 130,000 acres with "supplemental" water on the Payette-Boise project. In short, the Reclamation Service bragged that it had generated new wealth and prosperity for Idaho's reclamation project settlers.[40]

Still more land and water awaited takers in southern Idaho in 1925. A USRS engineer identified 1,437,000 acres within the Snake River Plain that could be irrigated at reasonable costs; and the Idaho commissioner of reclamation believed that the water spilling over Milner Dam into the Snake River canyon could be used for the purpose. However, few were willing in these times to give Carey Act reclamation a second chance. Although it had tapped southern Idaho's land and water resources to bring economic betterment to the region, it did not achieve as much as expected. So southern Idahoans acquired new dams and reservoirs in other ways. Today, facilities at Deadwood, Black Canyon, American Falls, and Palisades are some of the monuments to success in post-Carey Act times.[41]

This article originally appeared in *Pacific Northwest Quarterly* 78 (October 1987). Reprinted with permission.

Notes

1. Benjamin Horace Hibbard, *A History of the Public Land Policies* (1924; rpt. Madison: University of Wisconsin Press, 1965), 454.

2. *Caldwell (ID) Tribune*, July 27, 1907.

3. *Twin Falls (ID) News*, October 27, 1905; Merrill D. Beal and Merle W. Wells, *History of Idaho*, 3 vols. (New York: Lewis Historical Publishing, 1959), 2:139; *Biennial Report of the State Engineer to the Governor of Idaho for the Years 1899–1900* (Boise, n.d.), 7–8, 10 (hereafter cited as *Report of the State Engineer* with appropriate years); D. W. Ross, "Work of the Reclamation Service in Idaho," *Pacific Monthly* 16 (1906), 313.

4. [D. W. Ross?], "Irrigation Investigation in Idaho in 1902 and 1903 (typescript, December 1, 1903), and *idem*, "Summary of Investigations in Idaho" (typescript, January 26, 1904), Bureau of Reclamation (BR) Records, RG 115, File 123, Box 32/91728, Acc. No. 65-A-436, Operations Branch, Federal Archives and Records Center at Denver (FARC-Den.).

5. Fred A. Voigt to Charles H. Deppe, March 26, 1910, Twin Falls Land and Water Company Papers, Idaho State Historical Society (ISHS), Boise; Idaho state engineer to E. A. Keyes, October 7, 1910, Idaho Reclamation (IR) Records, Collection AR-20, Idaho State Archives (ISA), Boise; *Rupert (ID) Pioneer-Record*, August 12, 1909 (quotes).

6. For summaries of rules governing projects, see Mikel H. Williams, *The History of Development and Current Status of the Carey Act in Idaho* (Boise: Idaho Department of Reclamation, 1970), 1–10.

7. Ibid., 16–18, 52–53; Patricia Lyn Scott, "Idaho and the Carey Act, 1894–1930," (master's thesis, University of Utah, 1983), 101–32; Leo Sweet to Wayne Darlington, December 1, 1903, and D. G. Martin to Keyes, December 13, 1910, IR Records; *Report of the State Engineer, 1899–1900*, 22, 72–74.

8. "History of the Minidoka Project, Idaho, to 1912," vol. 1 (typescript), 4–8, 27; Project Histories and Reports of Reclamation Bureau Projects: Minidoka Project Idaho, 1910–1919, BR Records, RG 115, M96, Roll 96, National Archives; *Caldwell Tribune*, December 19, 26, 1903; Neil H. Carlton, "A History of the Development of the Boise Irrigation Project" (master's thesis, Brigham Young University, 1969), 47, 51–53; F. H. Newell to Darlington, March 10, April 4, 1904, IR Records.

9. *Chicago Tribune*, November 20, 27, 1909, January 17, 1910; *Chicago Record-Herald*, December 14, 1908, February 9, 1909; *Twin Falls News*, January 6, 1910; Henry F. Cope, "Mak-

ing Gardens Out of Lava Dust," *World To-Day* 10 (1906), 621–28; William E. Smythe, *The Conquest of Arid America*, rev. ed. (1905; rpt. Seattle: University of Washington Press, 1969) 186, 196.

10. Scott, 116; *Twin Falls News*, April 30, 1909; *Farm, Stock, and Home*, September[?] 1909, quoted in *Rupert Pioneer-Record*, September 23, 1909 (1st quote); J. E. Bangs, "Irrigation Securities vs. Other Investments: Pertinent Reasons for High Standing of Irrigation Bonds," *Bonds and Mortgages* [Chicago], October 1910 (copy in IR Records) (2nd quote).

11. William Darrell Gertsch, "The Upper Snake River Project: A Historical Study of Reclamation and Regional Development, 1890–1930," (PhD diss., University of Washington, 1974), 60–62; *Shoshone (ID) Journal*, November 22, 1907; *Twin Falls News*, June 7, 1907, May 21, 1909; James Stephenson Jr., *Irrigation in Idaho*, U.S. Department of Agriculture, Office of Experiment Stations Bulletin No. 216 (Washington, DC, 1909), 49–50; Bruce L. Schmalz, "Headgates and Headaches," *Idaho Yesterdays* 9 (Winter 1965–66), 22–25.

12. The Idaho Land Department calculated the cost of the South Side irrigation system at $3,394,676; Buhl claimed in 1908 that he spent $3,500,000 on the system (*Thirteenth Biennial Report of the State Land Department of the State of Idaho, 1914–1916* [Boise, 1916]. n.p.); Louise Morgan Sill, "The Largest Irrigated Tract in the World," *Harper's Weekly* 52 (October 17, 1908), 11 (quote).

13. Voigt to American Falls Canal and Power Company, November 28, 1908, Twin Falls Land and Water Company Papers (quote); *Twin Falls News*, August 23, 1907, December 10, 1909.

14. Byron Hunter and Samuel B. Nuckols, *An Economic Study of Irrigated Farming in Twin Falls County, Idaho*, U.S. Department of Agriculture Bulletin No. 1421 (Washington, DC, 1926), 7; Robert McCollum to Addison T. Smith, January 30, 1905, Addison T. Smith Papers, ISHS (1st quote); secretary of Twin Falls Land and Water Company to Mrs. A. M. Merritt, August 3, 1906, Twin Falls Land and Water Company Papers (2d quote); Katharine Coman, "Some Unsettled Problems of Irrigation," *American Economic Review*, vol. 1 (March 1911), 15; *Rupert Pioneer-Record*, March 18, 25, 1909 (knockers).

15. Voigt to J. H. Brady, July 13, 1910, and J. P. Whitla to Idaho Board of Land Commissioners (IBLC), July 22, 1910, James H. Brady Papers, ISA.

16. *Pocatello (ID) Tribune*, January 14, 1909.

17. *Reclamation Record*, vol. 5 (1914), 56 (1st quote); Ellen C. Talbot to Smith, July 25, 1908, Smith Papers (2d quote); Arthur Ruhl, "Those Who Wait: The Reclamation Service vs. Private Enterprise," *Collier's*, vol. 44 (January 22, 1910), 22–23, 26–27.

18. *Twin Falls News*, December 5, 1905 (1st quote); A. E. Robinson to J. Arthur Eddy, August 11, 1911, IR Records (2d quote).

19. Heber Q. Hale to Brady, August 23, 1910, D. R. Niver to Brady, March 24, 1910, Brady Papers; S. D. Taylor to F. P. King, February 16, 1914, IR Records; Schmalz, 22–23; *Shoshone Journal*, November 27, 1907 (quote); A. M. [Alexander McPherson] to J. M. Whittaker, November 9, 1909, Kings Hill Extension Irrigation Company Papers, ISHS.

20. "History of the Carey Act Project Originally Known as the West End Twin Falls Irrigation Project" (typescript, September 10, 1919), David W. Davis Papers, ISA; Frank K. Welles, *The Story of a Carey Act Investment* (Salem, OR: N.p. 1914), copy in Moses Alexander Papers, ISA.

21. *Twin Falls News*, December 1, 1905; Beal and Wells, II, 145; *Report of the State Engineer, 1907–1908*, p. 184.

22. *Report of the State Engineer, 1907–1908*, p. 184 (1st quote); Brady to W. B. DeJarnatt, April 30, 1909, Brady Papers (last quote); Hugh T. Lovin, "A 'New West' Reclamation Tragedy: The Twin Falls-Oakley Project in Idaho, 1908–1931," *Arizona and the West* 20 (1978), 5–24; Lovin, "Free Enterprise and Large-Scale Reclamation on the Twin Falls-North Side Tract, 1907–1930," *Idaho Yesterdays* 29 (Spring 1965), 2–14; Lovin, "How Not to Run a Carey Act Project: The Twin Falls-Salmon Falls Creek Tract, 1904–1922," *Idaho Yesterdays* 30 (Fall 1986), 9–15, 18–24; W. H. Rosecrans to James Stephenson, April 20, June 1, 1908, IR Records; "Irrigating the Country between Boise and Mountain Home" (typescript, n.d.), 1–2, Davis Papers; *Twin Falls News*, September 18, 1908, August 10, 1911, September 19,1912, January 8, 1913.

23. "List of Segregated State and Other Lands, under Carey Act Projects in Idaho" (typescript, [1907]), IR Records; Frank Gooding to Burton L. French, April 16, 1908, Burton L. French Papers, Miami University Library, Oxford, Ohio.

24. IBLC Minutes, March 14, June 22, August 10, 1908, IR Records; *Twin Falls Times*, February 17, 1910; Archibald C. Milner to Chester C. Baird, October 6, 1911, Twin Falls Land and Water Company Papers; Milner to D. W. Davis, December 11, 1918, Davis Papers; *Twin Falls News*, April 18, 1912.

25. *Report of the State Engineer, 1915–16*, pp. 8–12; J. H. Smith to F. S. Minor, November 8, 1916, and Taylor to King, February 16, 1917 (1st quote), IR Records; *Thirteenth Biennial Report of the State Land Department*, n.p.; *North Side News* (Jerome, ID), May 26, 1910 (last quote).

26. *Report of the State Engineer, 1911–12*, pp. 56–57; [G. B. Archibald], "General Report on The Kings Hill Project" (typescript, n.d.) 106–36, Project Histories and Report of Reclamation Bureau Projects: King Hill Project, Idaho 1911–1926, BR Records, RG 115, M96, Roll 61, National Archives.

27. *Boise (ID) Citizen*, October 14, 1910, cited in Randall R. Howard, "Irrigation Frauds in Ten States," *Technical World Magazine* 17 (July 1912), 514 (quotes); *Twin Falls Times*, October 21, 1910; Alexander to Franklin K. Lane, n.d., Alexander Papers.

28. For examples, see A. M. Bowen to Alexander, December 12, 1914, Alexander Papers; Williams, 80–81.

29. IBLC Minutes, August 13, 1917, and J. H. Smith to IBLC, January 24, 1918, IR Records; T. A. Larson, *History of Wyoming* (Lincoln: University of Nebraska Press, 1965), 358.

30. Niver to Brady, June 9, 1910, Brady Papers; "Outline on Report on King's [sic] Hill Irrigation Companies" (typescript, n.d.), Kings Hill Extension Irrigation Company Papers; Alexander to R. M. McCracken, February 4, 1917, Alexander Papers; "Explaining the Pittsburg[h] Crash," *Literary Digest* 47 (July 19, 1913), 85.

31. W. W. Vernon to Robinson, April 5, 1912, IR Records (quote); Ray Palmer Teele, *Irrigation in the United States: A Discussion of Its Legal, Economic and Financial Aspects* (New York: D. Appleton & Co., 1915), 133–34.

32. IBLC Minutes, January 29–February 3, 1916, Alexander Papers; J. E. Clinton to Joseph Otis, June 27, 1913, John M. Haines Papers, IDA; IBLC Minutes, May 29, 1917, IR Records (quotes).

33. R. P. Teele, "The Financing of Non-Governmental Irrigation Enterprises," *Journal of Land and Public Utility Economics* 2 (1926), 437–38; phrase quoted in C. C. Moore to True Webber Co., August 16, 1924, and R. E. Shepherd to Moore, January 27, 1925, Charles C. Moore Papers, ISA.

34. Taylor to King, February 16, 1914, state engineer to Frank R. Spier, November 27, 1912, and IBLC Minutes, December 17, 1917, IR Records; Haines to Lane, January 15, December 7, 1914, Lane to Haines, February 13, 1914, Haines Papers.

35. Gertsch, 88–91; A. L. Gilmore to Alexander, April 16, 1916, Alexander Papers (quote).

36. Alexander to William H. King, December 13, 1917 (1st quote), and to Lane, November 30, 1916 (2d quote), Alexander Papers; *Idaho Statesman* (Boise), February 9, 1918 (last quote).

37. IBLC Minutes, August 13, 1917, IR Records; Twin Falls Salmon River Land and Water Company v. M. Alexander et al., Idaho Attorney General Case File X-326, ISA; Lovin, "'New West' Reclamation Tragedy," 16–20; *Arco (ID) Advertiser*, August 20, 1920.

38. Beal and Wells, 2:33, 239; John F. Nugent to Ernest G. Eagleson, August 13, 1919 (quotes), and Eagleson to George B. Hill, January 22, 1920, Ernest G. Eagleson Papers, ISHS.

39. Swendsen to Moore, August 16, 1924, Moore Papers; Minutes of Governor's "Cabinet," May 10, 1919, pp. 5–6, Davis Papers; S. D. Taylor, *Carey Act Projects: Report on the Industrial Development of Idaho, Accompanied by the Reclamation of Desert Lands by Actual Irrigation under the Carey Act* (Boise: N.p., 1913), 5; *Thirteenth Biennial Report of the State Land Department*, n.p.

40. Department of the Interior, U.S. Reclamation Service, "Idaho: Boise Project and Related Features" (mimeographed, 1920), 12 (copy in French Papers).

41. Harold Conkling to chief of construction, January 26, 1920, BR Records, RG 115, Engineering Correspondence File 201i, Box 787, National Archives, FARC-Den.; Swendsen to Moore, October 8, 1925, Moore Papers.

PART III

Projects: Failures and Successes

Steam shovel digging the main North Side Canal from Milner Dam, ca. 1905. *Twin Falls Public Library, Clarence E. Bisbee Collection, PC-2084*

CHAPTER 5

Free Enterprise and Large-Scale Reclamation on the Twin Falls North Side Tract, 1907–1930

Citizens seeking free or inexpensive western land on which to settle have been beneficiaries of a number of federal laws. As early as 1790 Secretary of the Treasury Alexander Hamilton advocated the sale of public lands at extremely low cost—and in units too small to encourage land speculation—in order to encourage settlement of the country to the west of the new nation's population concentration. By the Civil War, most of the public domain had been taken up; of the billion or so acres remaining, only about a third was arable land. The Homestead Act of 1862 served as the first major vehicle for disposing of that third. It was followed by such varied legislation as the Timber Culture Act (1873), the Desert Land Act (1877), and the Dawes Severalty Act of 1887 (which allowed white settlers to claim lands on Indian reservations), as well as railroad land grants.[1]

Many provisions of the federal land laws came to encourage fraud and abet speculation in land instead of promoting real agricultural reclamation of the arid western regions. In response to such problems—and to encourage investment in the technology necessary to bring about large-scale irrigation—Congress passed the Carey Act of August 18, 1894. The new law invited private entrepreneurs to finish the job on 1,000,000 acres in each state that possessed potentially irrigable land, subject only to state supervision. In return for erecting essential irrigation structures, these entrepreneurs could sell water to settlers, profits being the excess of settlers' payments over construction costs on each reclamation project.

In the next three decades, private capital accepted the challenge of reclaiming numerous Rocky Mountain valleys and even large portions of several intermontane plateaus; their busiest laboratories were in Colorado, Idaho, and Wyoming.[2] There, though the new Carey Act projects tended to be modest in size and capitalization (especially as compared to settler expectations), corporations also tackled more difficult conditions and infinitely greater hazards incident to large-scale reclamation. Typical of the latter enterprises, all of them sharing the same characteristics of risk to entrepreneurs and investors and allure in their possibilities of success and riches, was the development of Idaho's Twin Falls North Side Project.[3]

The Idaho Board of Land Commissioners, commonly called the Land Board, authorized the state's first Carey Act project in 1895 and theoretically stood ready to approve more; but the board, which would be responsible until 1919 for supervising such projects, twiddled its thumbs for another half-decade. Until the depression of the 1890s subsided, entrepreneurs declined to risk capital on irrigation systems. Finally, on October 12, 1900, industrialists Frank Buhl and Peter Kimberly of Sharon, Pennsylvania, took the plunge, contracting with the board to build irrigation works on the 244,000-acre Twin Falls South Side Project in Twin Falls and Cassia Counties. This tract, with fertile land suited to inexpensive gravity irrigation from the Snake River, nearly overnight attracted settlers from afar and soon received national notice. Its admirers included trusted agricultural scientists; the project's promoters even claimed an endorsement from Liberty Hyde Bailey, Cornell University scientist and head of President Theodore Roosevelt's Country Life Commission; and William Jennings Bryan characterized the project as a "garden."[4]

Such publicity and quick profits from selling water—exaggerated by journalists—resulted in others deluging the Land Board with proposals for fifty Carey Act projects for which they planned irrigation works costing $68,479,015.[5] Several of this newest batch of applicants proposed projects even more ambitious than the South Side tract. The promise of largest investment came from James S.

and William S. Kuhn, Pittsburgh commercial and investment bankers, coal and traction car enterprisers, and principal stockholders in the American Water Works and Guarantee Company of Pittsburgh. William Kuhn, who took charge of the brothers' western affairs, at the same time acquired the defunct Central Irrigation District project in California's Sacramento Valley. In 1906 and 1907, he also gained control of small companies that held permits to water power sites and operated hydroelectric plants in Idaho. Kuhn anticipated profits from the sale of electricity to townsmen and farmers when reclamation tracts filled the Snake River plateau. Kuhn's best power markets might well include copper smelters in northern Nevada, should industrialist Solomon Guggenheim pursue his reported interest in tapping copper deposits in that state.[6]

Among Kuhn's requests to the Idaho Land Board was one for authority to reclaim about 350,000 acres adjacent to the thriving South Side Project. His proposal covered three separate tracts, each of which he christened a "Twin Falls" project because journalists invoked this name as a symbol of successful reclamation. The 185,000-acre Twin Falls North Side tract, largest of the three, was potentially the most lucrative as well even though Kuhn first had to purchase rights that Frank Buhl and his associates had staked earlier. Kuhn foresaw more than sales of water to farmers. He would market townsite lots and sell electricity from his adjacent Snake River power plants, and he intended to generate revenues from users of his Idaho and Southern Railway, an interurban electric railroad servicing the North Side region. Kuhn described his North Side plan: "to force the use of electricity by low rates and good service" while creating "the greatest irrigated tract in the world as an electrical irrigation tract."[7]

Altogether Kuhn planned investments of $30,000,000 in the West, the bulk of such capital to be derived from marketing stocks and bonds issued in the name of his Twin Falls North Side Land and Water Company and several other self-established corporations for his western enterprises. In order to reassure investors in these companies, Kuhn linked them to the American Water Works and Guarantee Company of Pittsburgh—which he and his brother dominated—so that it functioned as guarantor of all bonds issued by Kuhn's western companies. The guarantor assumed responsibility "by assignment, or agreement," for the "agreements and contracts" of these western companies. Such conditions impressed bond-market investors nationwide. They quickly subscribed the first $11,700,000 that Kuhn wished to raise, and he earmarked most of the funds for constructing dams, reservoirs, and other irrigation structures on his California and Idaho reclamation tracts.[8]

The Idaho Land Board reviewed Kuhn's reclamation proposals uncritically. Board members were elated that Kuhn wanted to become the state's foremost reclamation financier—replacing Kimberly, who had died recently, and Kimberly's partner Frank Buhl, who now doubted that reclamation projects rewarded financiers sufficiently for their risk-taking. Buhl's agents had privately advised the Land Board that Twin Falls South Side investors earned profits of $7.00 per acre, an amount not "commensurate with the investment."[9] Kuhn then appeared before the board and, always gregarious and genial, charmed the members. The board had already checked the financial ratings of Kuhn's eastern corporations and was awed by his wealth and reputation in business circles for acumen and fairness.[10] Nor, state engineer James Stephenson Jr. advised the board, were there serious technical objections to Kuhn's reclamation schemes; later Stephenson lauded the entrepreneur's methods because Kuhn "condenses within three or four years the progress which, under ordinary circumstances, requires twenty years."[11]

The Land Board's romance with Kuhn, after first blooming in 1907, lasted for another six years. The board granted his request for an 185,000-acre North Side project, increased the project to approximately 235,000 acres in 1908 at Kuhn's request, and added another 50,196 acres in 1910.[12] Kuhn and the board signed contracts in 1907 and 1909 that incorporated Kuhn's plans for irrigating the project. Besides diverting water from the Snake River at Milner Dam, Kuhn planned to store 80,000 acre-feet in "on-site" reservoirs within the project, and he was obliged to build canals, laterals, weirs, and head gates serving each of three project sectors to be opened in sequence to settlers. The Land Board agreed that Kuhn could assess his Twin Falls North Side Land and Water Company settlers substantially more than South Side Project farmers paid for water; the North Side assessments ranged from $30 to $45 per acre, depending upon the project sector. Kuhn calculated handsome profits from such collections because his blueprints showed outlays for North Side irrigation works not exceeding $2,500,000, and Kuhn's profit margin still seemed acceptable when $500,000 in cost overruns loomed in 1909.[13]

Also without careful scrutiny, the Land Board accepted Kuhn's engineering blueprints, documents based on premises and irrigation concepts crucial to project success. From Milner Dam, coterminous with the project's south perimeter, Kuhn would divert at least 1,800 second-feet of water to North Side lands. Small as this water supply was—one-fifth of what North Side irrigators later consumed yearly—the Land Board accepted Kuhn's arguments that such quantities sufficed momentarily, and the board trusted him to procure more water as settlers needed it. Kuhn ultimately filed claims to another 22,000 second-feet of Snake River water, but he never developed his own upstream reservoir sites in order to divert the water to the North Side tract. Instead he purchased annually from 1907 to 1910 another 140,000 acre-feet of water from the U.S. Reclamation Service. This water, impounded at Jackson Lake, Wyoming, was available temporarily because the agency's Minidoka Project could spare the water.[14]

The Land Board's acceptance of Kuhn's calculations on water sufficiency raised eyebrows outside official circles in the state. For so large a project, these observers noted, the board had agreed to a rather modest water supply, and access to part of that was tenuous because of unanswered questions about the Reclamation Service's authority to sell Jackson Lake water to any non-federal entity. The board defended itself, holding that in cases of water shortages North Side irrigators must adhere to "good public policy" in the state, which meant strict water conservation. Furthermore, the board reminded its critics, North Side irrigators possessed no claim to specific amounts of water; the fine print in the 1907 contract between Kuhn and the board instead defined their entitlements as only such amounts "as the conditions of the crop and weather may determine" and "as will best serve to protect the interests of all water users."[15]

Kuhn's plans were challenged on other grounds as well. Observers, including a number of Statehouse officials, were suspicious of Kuhn's scheme to transfer water from Milner Dam during the non-irrigating season, store it in a natural coulee near Sugar Loaf Butte and several smaller depressions in the terrain, and from these on-site reservoirs deliver water to farmers in the summer. Admittedly Kuhn had selected an economical procedure, substituting cheap onsite structures for expensive Snake River dams that impounded water in excess of Milner Dam's storage capacity. But critics reminded Kuhn and his allies of the porous quality of volcanic strata beneath the North Side region, and they claimed that Kuhn's fragmentary geological data could not demonstrate that North Side coulees could ever retain water. Comparing North Side reservoirs to similar structures that had failed elsewhere on the Snake River Plain, the same detractors pointed out that one Payette-Boise Project reservoir had wasted water copiously, and Reclamation Service engineers might never be able to repair the structure satisfactorily. For the moment, however, Kuhn had the last word on the matters. He invoked the ideas of western reclamation pioneer John Wesley Powell, who had long ago recommended small on-site reservoirs instead of massive dams across rivers, and the Land Board predictably sided with this invoking of Powell's expertise.[16]

Denigration of Kuhn's engineering continued, but his prestige in Land Board eyes, recognition of his contribution to the state's post-1900 reclamation boom, and his explanations of Powell's theories—which many Idaho newspapers dutifully published—finally reduced the criticism of the North Side Project. Many Idahoans, in fact, concluded that the remaining detractors were simply hardcore believers in reclamation at federal hands working in league with certain old-line stock raisers whose self-serving motives had become obvious. The latter seemed unlikely ever willingly to surrender a drop of water or an acre of grazing land to intruding reclamationists.

Kuhn's critics persisted, particularly in deploring the Land Board's agreement with Kuhn and in predicting disaster ahead on the North Side tract. Their concerns finally shifted the focus of debate to questions about the state's wisdom in trusting free enterprise to reclaim the region's arid expanses. The shift proved a costly mistake for Kuhn-Land Board critics. Kuhn's Idaho admirers and his out-of-state supporters pointed out that "private capital" asked only to make "manless domain" succumb to "the energy and determination of empire builders." Quickly endorsing this view, Idaho state officials called for reclamation by private hands of the entire Twin Falls country, where they estimated another 1,000,000 acres awaited reclaimers.[17]

Kuhn's admirers, more than a match for opponents, then exploited their lead. They noted that his enterprises had always flourished, and fair play dictated that he not be prejudged otherwise. They also contrasted the achievements of private capital with those of the Reclamation Service. Such comparisons exploited old prejudices and hurts in a state where, whatever the residents' respect for President Theodore Roosevelt, the acts of the President's Reclamation Service administrators were controversial. In particular, Kuhn's supporters stressed certain contrasts between the state's Carey Act showpiece, the Twin Falls South Side Project, and the federal Minidoka Project. From what they called a fair comparison because soil, climate, and other conditions were virtually identical on the two projects, Kuhn's forces always adduced evidence of inferior results on the federal installation, and National Woolgrowers' Association president Fred Gooding proclaimed that intermontane reclamation provably belonged to the "beneficent control of private capital directed by the state[s]." Public attacks on federal reclaimers at Minidoka finally became so commonplace that Reclamation Service Director Frederick W. Newell protested against the sport. To no avail, he complained that these persistently invidious allusions to Minidoka were unwarranted, had "destructive effects" on the federal project, and were in any event irrelevant to debates between Kuhn and his opponents.[18]

Unable to sidetrack the North Side Project as Kuhn proposed to develop it, his enemies dared him to succeed, and at first it appeared that he would confound them. From 1907 to 1912, he placed the project on firm financial footing while protecting his interests and those of North Side Company bondholders and incoming land entrymen. He readily raised capital as it was required, each time floating issues of North Side Company 5 percent gold bonds, and Kuhn's competitors complained about investors preferring such bonds to other irrigation securities.[19] Furthermore, neither episodes in "Blue May" of 1910, which weakened confidence in western irrigation securities, nor the market's collapse a year later following the bankruptcy of two Chicago brokerage houses that handled irrigation securities, troubled Kuhn. He persuaded investors that North Side land values would skyrocket. Some bought land and water rights as well as bonds, paying with cash or, for the former, with promissory notes maturing in 1919; others replenished North Side coffers by purchasing townsite lots for homes and commercial ventures.[20]

At the same time, Kuhn's agents lured settlers to project farms. Their favorite advertising medium was exhibits at urban "land expositions" that agents had sometimes masterminded themselves. There Kuhn's representatives preached, often to relatively affluent sons of the Middle Border [Midwest], that the Promised Land lay in Idaho, and they emphasized to potential entrymen the North Side Project's amenities—for example, irrigation works that Kuhn planned to have ready ahead of the Land Board's deadlines for water delivery. Kuhn proudly displayed his irrigation works, then under construction, to visitors inspecting the North Side tract. Enough speculators and bona fide settlers were interested that the groups purchased irrigation water for 146,890 acres and, in a single land drawing October 1, 1907, purchased land and water rights worth $2,040,000. The most thrifty paid in cash, though Kuhn offered an installment plan, and thus evaded his stiff interest rates.[21] These first entrymen settled on their land in 1907 and 1908—many of them counting on cash reserves sufficient to sustain them until 1909, when water would reach their lands and make their farms productive.

Kuhn also reassured North Side entrymen by announcing his own "Gospel of Wealth," which was somewhat akin to what Andrew Carnegie had prescribed in 1889 for capitalists: amassing profits while offering settlers his "helping hand" and "warm heart." He established demonstration farms where agricultural scientists showed newcomers the best crops, promoted dairying, and displayed irrigation techniques to neophytes. Kuhn loaned money to farmers for purchasing dairy cattle and seeds for planting, and he donated both land and money to establish schools, parks, and public buildings in four project towns.[22]

Kuhn's agents claimed that these practices, said by one subordinate to be followed when "good business" permitted, generated prosperity on the project within five years after settlers first arrived in 1907. In several villages, commerce hummed in 1912, providing goods and services to five hundred farm families and the townspeople. In the same year, Kuhn's representatives calculated that farmers harvested alfalfa, potatoes, and fruit from 32,000 acres and earned additional incomes from dairy herds and sheep flocks.[23]

Equally good, it seemed, were the economic fortunes of Kuhn and of North Side Company bondholders, who had risked several million dollars. Company files bulged with water sales contracts with entrymen; sales of water for additional land continued, albeit not as briskly as Kuhn desired; and the Idaho Land Board and state courts had protected these instruments of indebtedness. In upholding the contracts, judges permitted the company to foreclose on water rights of settlers falling in arrears and ruled that errant settlers could lose their land to company foreclosure as soon as settlers had received their federal land patents. The Land Board was even more adamant. Under penalty of foreclosure, the board reminded landholders, they must abide by contracts after settlers had received water for "one irrigation season." When some entrymen objected, asking the board to sanction nonpayment until Kuhn redressed certain grievances, the board ordered these settlers to pay lest [their] non-payment compromise "the integrity of the bonds upon which the money was raised to build the project."[24]

Obviously, not everyone had prospered in the project's first five years. Only a fraction of the entrymen still trusted Kuhn and described themselves as reasonably contented. One group of settlers, who occupied the lands nearest Milner Dam where water was plentiful, had prospered enough that some believed "Easy Street" was just around the corner. Conversely, farmers situated in the central and western portions of the tract had endured water shortages since 1909, the shortfalls stemming from problems in the main on-site storage reservoir at Sugar Loaf Butte. At the worst, farmers in these sectors claimed impoverishment. One settler described his plight: "I had a nice little bunch of money when I came here. Today I have just 50 cents left in the house and hardly enough to eat."[25] Nor did these grumbling entrymen expect quick relief, largely because the romance between Kuhn and the Land Board seemed as torrid as ever. Stopping just short of accusing the board and Kuhn of conspiracy, entrymen blamed the board for tolerating a "comedy of errors," overlooking Kuhn's greed, and shirking its duty to ensure that settlers were "protected and shielded."[26]

Under such circumstances, a settler faction coalesced to take control of the local Canal Users' Association, which had since 1909 represented the entrymen but maintained cordial relations with Kuhn and his resident agents. The rebel group took over the association and shored up the organization in anticipation of punishing Kuhn, whom Canal Users' Association president L. T. Alexander characterized as an "oppressor." Alexander commented:

> For every dollar the Kuhns have poured into Idaho, the settlers have poured several dollars in money besides four or five best years of their life...[Anyway] it wasn't Kuhn money. They sold their bonds to widows, orphans, estates, [and] poor laboring people who had saved a few hundred dollars for investment.[27]

Attorneys counseled the rebels to avoid "anarchistic methods," so they rallied settlers at public "indignation" meetings and eventually claimed support from "nineteen-twentieths" of all North Side farmers. The dissidents then petitioned state officials to impose financial penalties on Kuhn, and they finally announced their ultimate demands. They insisted that the Land Board terminate Kuhn's authority on the North Side project, cancel all state contracts with him, and compel forfeiture of certain performance bonds that Kuhn had deposited with the board in 1907. Land Board members reluctantly discussed these draconian measures on October 23, 1912, reached no decision that day, and rejected the proposals on January 2, 1913.[28]

Meanwhile, the controversy had spread beyond the project and Land Board realms, and rebels counted on enlisting new support in pressuring the Land Board to discipline Kuhn. Their "big row" with Kuhn interested journalists, even out-of-state newspapermen, and the *New York Times* recounted the settlers' side of the fight on October 6, 1912. In the 1910 and 1912 state election campaigns, the rhetoric of both major parties (and Bull Moosers in the latter election year) echoed North Side dissidents' views. But such political commentary scarcely served rebel purposes. As observers noted, these politicians had vied for reclamation-tract votes when their candidates were locked in close races for office; besides, many Idahoans discounted such "buncombe" [bunkum], recognizing that in Idaho politics candidates for public office usually positioned themselves with "the settler in his fight against the ditch companies and promoters."[29]

The North Side Project dissidents expected more stalling from Kuhn and the Land Board, but Kuhn surprised them. He disputed settlers' accusations, but he also had already attempted to atone for certain mistakes. He accepted responsibility for conditions at Sugar Loaf Butte, where, as in a sieve, impounded water wasted into basalt strata beneath the site; his agents opened negotiations with entrymen over reimbursement for crop losses that resulted in 1909 and 1910. He also reconsidered his plans for on-site reservoirs, deciding to abandon them and, with Land Board concurrence, effect arrangements with the Reclamation Service for equivalent water storage at Jackson Lake. He next negotiated with Secretary of the Interior Richard A. Ballinger, who at last agreed to share Jackson Lake water with the entrepreneur. However, Ballinger drove a hard bargain. He demanded that Kuhn contribute proportionately to financing a large concrete dam on Jackson Lake that would serve long-term water requirements of both the North Side and the

Minidoka project; and, before such a dam could be constructed, Ballinger compelled Kuhn to deposit at least three-fourths of his share of the funds. Furthermore, Ballinger conditioned the offer on passage of the Warren Act, legislation—stalled in a congressional committee—that authorized the Interior Department to sell federally impounded water to private irrigators.[30]

Kuhn finally despaired of any federal partnership when Congress delayed its approval of the Warren Act until 1911 and the Reclamation Service raised higher roadblocks by refusing to adjust Jackson Lake construction timetables to the urgency of North Side water needs. He went back to his concept of an on-site reservoir, deciding to keep the Wilson Lake Reservoir (which had always retained water) and build another one. Construction of the latter, Jerome Reservoir, was completed at a cost of $500,000.[31]

———♦———

Misfortune struck anew. Like its predecessor at Sugar Loaf Butte, the new reservoir wasted its water into cavities and cracks permeating the volcanic lava that underlay the coulee. Kuhn partly compensated for the losses, but his measures dismayed the entrymen. Over settler protests that he had resorted to a "subterfuge for cutting down the individual's water supply," Kuhn's agents spread the remaining water among farmers by enforcing a new water-rotation system on the tract. He also convinced the Reclamation Service to sell him certain Jackson Lake water that was still surplus to the needs of the Minidoka Project. Idaho farmers upstream salvaged a few North Side crops by contributing another 1,500 second-feet of water in 1911.[32]

Kuhn considered several ways to overcome the engineering problems that remained. Finally he heeded advice from the state engineer, his own engineers, and a panel of outside irrigation consultants, all of whom recommended that he simply patch the main leaks at Jerome Reservoir.[33] Repairs commenced as soon as practicable, while he mulled over his second thoughts about the wisdom of incurring the risks inherent in large-scale reclamation. He might have a way out: assuming that the repairs to Jerome Reservoir were successful, the language of Kuhn's contracts with the Land Board appeared to excuse him from more construction obligations or legal responsibility for procuring additional water for the North Side project. He therefore announced that his finishing touches at Jerome Reservoir "completed" reclamation of the tracts from their original arid condition and released him from his 1907 contract to store 80,000 acre-feet of water annually in on-site reservoirs. He also interpreted the Carey Act to his benefit, saying that North Side land holders were now entitled to patents on 150,563 acres; if his interpretation were upheld, he had strengthened his clout in enforcing water contracts with settlers because, once federal land patents were issued, his North Side Company could take possession of the land taken by delinquent water users. However, Kuhn's expectations were premature. The refurbished Jerome Reservoir failed again; after repairs, it retained in 1912 about one-half of the stored water for which engineers had designed the structure.[34]

For the first time, the Land Board systematically reviewed North Side Project problems rather than accepting Kuhn's theories and solutions. To his dismay, the board and its technical advisor, state engineer A. E. Robinson, concluded that North Side farmers required 130,000 acre-feet of water annually from on-site reservoirs or alternate sources. Kuhn balked at the requirement that he supply more than the 80,000 acre-feet of on-site storage attainable once his engineers corrected the Jerome Reservoir's troubles. Furthermore, the project possibly faced a worse water crisis. The Land Board worried that litigation pending in Idaho courts could deprive Kuhn of Snake River water to which he had long before staked claim and dedicated to North Side usage. Kuhn again disputed the board's conclusion. Finally, Robinson persuaded the board to compel Kuhn to place 130,000 acre-feet of water in storage, and the board unilaterally imposed the requirement when Kuhn objected.[35] Kuhn continued to protest; the board's precipitate acts were ethically dubious and probably, as Kuhn argued, breached his contracts with the board.

Nevertheless, Kuhn complied under duress with the board's orders that he substantially augment North Side Project water resources; and he cooperated when the board, after judging abandonment of leaky Jerome Reservoir inevitable, instructed him to procure long-term access to 170,000 acre-feet of Jackson Lake water annually. By Land Board accounting, such quantities sufficed for immediate North Side irrigation, compensated for probable losses of Snake River water at the hands of state courts, and placed North Side water supplies beyond Idaho courts' tampering. Kuhn then contracted with the Reclamation Service on February 26, 1913, purchasing 150,000 acre-feet of water for immediate delivery annually, and he agreed to absorb nearly all costs of raising Jackson Lake Dam so that the North Side Project would eventually have access to 322,000 acre-feet of water.[36] Water exceeding the 170,000 acre-feet minimum, as decreed by the Land Board, presumably ensured reclamation of another 100,000 unoccupied acres to which Kuhn intended to sell water rights.[37]

Although Kuhn disliked incurring extra expenses in complying with Land Board directives, his cooperation paid dividends in the form of release from more North Side responsibilities. Holding that North Side water requirements were now satisfied, the Land Board ruled that Kuhn's remaining obligations were covered in his contracts with the Reclamation Service. Also, as Kuhn requested, the board sought federal patents for settlers to 170,663 North Side acres. For all concerned, a board spokesman wrote, the tide had turned, and the project would "prosper as [much as] the [Twin Falls] South Side" tract.[38]

To speed construction of Jackson Lake Dam, which could not proceed until Kuhn deposited funds with Reclamation Service authorities, he offered investors new North Side Company bonds, and the first prospective subscribers inspected the North Side Project in May, 1913. However, before he could secure new capital, the Kuhn brothers' industrial and financial empire crumbled abruptly. Like dominoes falling in sequence, a Kuhn-controlled bank at Pittsburgh failed on July 7, 1913, triggering runs on other Kuhn banking institutions. These incidents alarmed stockholders in the Kuhn-dominated American Water Works and Guarantee Company, they forced the corporation into receivership, and the taint spread to corporations whose fates the Kuhn brothers had intertwined with the Water Works Company. Among the latter, the Twin Falls North Side Land and Water Company tottered.[39]

Many years elapsed before James and William Kuhn, their creditors, stockholders, and jurists unraveled this tangle.

As for the North Side Company, it was soon cast adrift. Receivers of the American Water Works and Guarantee Company reorganized the corporation in 1914, divested it of accountability for Kuhn's reclamation ventures, and satisfied bondholders of several Kuhn organized corporations in the West. Payment of $1,000,000, of which the majority was the pro-rata share of North Side Company bondholders, released the Water Works Company from all guaranty responsibilities for the several companies' bonds and other obligations.[40]

In Idaho, North Side settlers panicked, fearing loss of their land and water rights because of Kuhn's bankruptcy and his failure to procure Jackson Lake water for the tract, and the probable refusal of North Side Company bondholders to make investments to secure the water. However, Governor John Haines declared that the North Side Project would never perish for want of Wyoming water storage. Though Haines would hold North Side Company bondholders accountable for Kuhn's responsibilities, he demanded a new cooperative federalism to save troubled reclamation projects. In a pronouncement opposite to words that had long emanated from the statehouse, Haines stated that the "day of irrigation development on a large scale by private or corporate enterprise is a thing of the past," and federal intervention had become essential.[41] In particular, he proposed that the federal government adjust its terms to the ability of North Side Company bondholders to raise the same funds and otherwise meet conditions in Kuhn's 1913 contract with the Reclamation Service.

Interior Department officials responded sympathetically. Secretary of the Interior Franklin Lane publicly agreed to federal intercession to rescue "meritorious projects," while his subordinates told Haines that federal budgetary restraints merely required "joining of state and federal forces" in such a rescue. Haines rejected these suggestions of state financial commitments, and

Lane finally advised Haines privately that reclamation projects "requiring water storage or other expensive [irrigation] works" had federal priority. As for the North Side Project, its fortunes tied to Jackson Lake storage, Reclamation Service officials offered to supply water after making "proper arrangement[s]" with North Side Company bondholders or settler organizations.[42] And, as the Reclamation Service promised, the bureau relaxed its rules governing the agency's stewardship at Jackson Lake. As a temporary palliative, the Reclamation Service released its own water to the North Side Project in 1914 and 1915, trusting the good faith of North Side Company bondholders for eventual reimbursement.[43]

However, such arrangements deferred decisions on the more crucial matter of funding the project's share of dam construction costs at Jackson Lake. The Reclamation Service, refusing to bend again, insisted that either bondholders or settlers advance $500,000 for this purpose—the amount to which Kuhn had agreed before his bankruptcy. Bondholders demanded that settlers pay the bill; North Side entrymen refused; bondholders declined a second time to subscribe the money, and once more it appeared that long-term water shortages might doom reclamation of large sectors of the North Side tract. But Governor Moses Alexander and the Land Board finally tired in 1916 of these settler-bondholder controversies and pressures on them from both sides. Alexander and the board, however, refrained from imposing their wills except to release settlers from any new financial obligations.

Negotiations among Alexander, the Land Board, Reclamation Service officials and bondholders continued for months before a way out of the impasse over finances emerged. At last, all sides agreed to terms by which bondholders paid $150,000 in cash and deposited with the Reclamation Service an equal amount of commercial and surety bonds to guarantee payment for the new construction work. The agreement required, as additional protection for the Reclamation Service, that bondholders deposit $300,000 in surety bonds with the Idaho Land Board in 1916. This bonding scheme in effect granted North Side Company bondholders a breathing spell in which to raise $350,000. Under such conditions, the bondholders discharged their $447,000 obligation on completion of new construction at Jackson Lake in December, 1916—less than original cost estimates. The enlarged dam at the lake's outlet permitted the North Side Project to tap 315,000 acre-feet of water yearly and at last abandon Kuhn's undependable on-site reservoirs.[44]

The tract's worst woes—insufficient water and financial insecurity following Kuhn's bankruptcy—were addressed in 1916, and better days lay ahead. Generally North Side entrymen relaxed; they had larger and more secure water supplies at hand with few federal strings attached. In the next few years, the North Side Project experienced sustained agricultural growth. Because of "good and continued water service" in most years of economic stimulus from World War I, the cultivated area increased from 65,000 acres in 1915 to 120,380 acres in 1919. In the next decade, the tract's population rose to about 15,000 people. On 128,319 acres cultivated in 1930, farmers produced grain and alfalfa and growing quantities of potatoes, corn, and sugar beets. North Side farmers had also procured high-quality dairy cattle, and as early as 1923 the tract "exported" nearly $1,000,000 worth of dairy products.[45]

In this same period, bondholders discharged their last obligations as unwilling heirs to Kuhn's bargains. Again digging into their pocketbooks between 1916 and 1920, they corrected several deficiencies in project canals and laterals to the settlers' satisfaction. The state then ruled the project "complete," meaning that all Carey Act requirements for reclamation were met; in compliance with the act, management and operational control of the project passed to the North Side Canal Company. In the new canal company, North Side landowners voted in proportion to water entitlements appurtenant to their lands. When judged "complete," the project contained 185,000 acres of the 261,945 acres that Kuhn once was authorized to reclaim. The 185,000-acre project conformed to a federal district court ruling of September 10, 1917, mandating that project be no larger than its water resources could sustain.[46]

Another decade of friction lay ahead between settlers and the bondholders after 1916. Droughts twice depleted Jackson Lake water supplies, reducing the North Side Project's share of the

Surveyors' camp for the North Side canal project, 1909. Bisbee Photo. *Idaho State Historical Society, 73-221-804-a*

scarce water by 50 percent in 1919 and to a lesser extent in 1924. Many settlers argued that such conditions demonstrated that the project's 185,000 acres could not all be irrigated, and they accused North Side Company bondholders of "profiteering" from water contracts with settlers covering all 185,000 acres. Bondholders denied the charge, their spokesmen retorting that the bondholders anticipated $4,000,000 in losses on their investment.[47] Meanwhile, a few farmers urged a federal takeover of the project; the Reclamation Service discouraged the proposal, and the possibilities of federal control were so remote that bondholders ignored the proposal.[48]

The bondholders' agent, Russell Shepherd, then established permanent residence on the tract in the 1920s, emerging as both a civic leader and principal official of the North Side Canal Company, which now operated project irrigation works. After winning the confidence of many residents, he eventually persuaded them to his personal view, and the position of North Side Company bondholders, that farmers should purchase "supplementary" water supplies from the new American Falls Reservoir on the Snake River, even though farmer water costs already averaged about $42 per acre.[49] At this point, moreover, financial aid from the bondholders could not be expected. Shepherd constantly reminded the settlers that state officials and jurists had determined that only 185,000 acres of the original 261,945-acre tract could be satisfactorily reclaimed, and bondholders had incurred sufficient losses by their inability to sell water for nearly 77,000 acres. This loss amounted to about $3,000,000.

Even Shepherd's influence and leadership in North Side affairs could not remove some of the settler-bondholder rancor inherent in the debtor-creditor relationships between the two groups. Bondholders held water sales contracts that most settlers had not redeemed fully. Shepherd calculated in 1923 that payments of principal, which bondholders had deferred during bad times of the 1910s and the farm depression of 1921–22, plus accrued interest on these debts, could escalate many settler bills to $80 per acre. Nor did settlers dare try to evade their obligations. Bondholders had established ironclad liens on settlers' farms and water rights until debts were discharged. Over settler opposition, the Land Board permitted bondholders to extract liens in 1914 by making them the price of reprieves on settler water payments until March 1, 1919. Likewise, bondholders deferred settler payments during the 1921–22 depression only when farmers consented to liens on their land and water.[50]

Passage of time and debt redemptions removed the last bondholder-settler friction, water deficiencies were finally remedied at the settlers' expense, and the 185,000-acre North Side Project became more prosperous by any standard of economic measurement. Large-scale reclamation had been achieved, and free enterprise's role in transforming these tens of thousands of acres from arid wastelands should not be minimized. Like similar Rocky Mountain region projects, the North Side undertaking was initiated with unrealistic expectations and often unequivocal *laissez faire* postulations. Kuhn put some of his own funds and his entrepreneurial reputation on the line; North Side Company bondholders risked about $6,300,000 and finally recovered approximately half of that amount. But all had accepted the hazards in return for chances of earning profits. The adverse consequences scarcely gainsay capitalism's initiative and the importance of investor dollars to the ultimate results of large-scale reclamation of lands for which neither federal nor large amounts of private capital funding was available. Perhaps, as one admirer of Carey Act reclamation methods argued in the 1920s, it had been a "sound principle" of public policy for governments to encourage "capital" to seek western outlets such as the North Side Project.[51] At least, many individuals and the public weal are better off today because of free enterprise's contribution to the North Side Project and its successful cousins scattered around the Rocky Mountain region.

This article originally appeared in *Idaho Yesterdays* 29, no. 1 (Spring 1985). Reprinted with permission.

Notes

1. Probably the best available discussion of the history of the public domain is Roy M. Robbins, *Our Landed Heritage: The Public Domain 1776–1936* (Lincoln: University of Nebraska Press Bison Books, 1962), from which this brief discussion was drawn.

2. Because of demands in three states for more Carey Act projects than the existing land allocations could accommodate, Congress granted an extra 1,000,000 acres each to Colorado and Wyoming and an additional 2,000,000 acres in Idaho. Benjamin Horace Hibbard, *A History of the Public Land Policies* (reprint ed.; Madison: University of Wisconsin Press, 1965 [1924], 436–37.

3. A satisfactory history of Carey Act reclamation projects remains to be written. For examples of different genres of writing in which parts of the story are recounted, see Hibbard, *Public Land Policies*, 434–39, 452–55; "Reclamation of the Arid Lands," Chapter 22 in Paul W. Gates with Robert W. Swenson, *History of Public Land Law Development* (Washington, DC: Public Land Law Review Commission, 1968 reprint ed., New York: Arno Press, 1979), 635–98; Samuel P. Hays, *Conservation and the Gospel of Efficiency: The Progressive Conservation Movement, 1890–1920* (Cambridge: Harvard University Press, 1959); Dorothy Lampen, *Economic and Social Aspects of Federal Reclamation*, Johns Hopkins University Studies in Historical and Political Science Series 48, no. 1 (Baltimore: Johns Hopkins University Press, 1930); Stanley Roland Davison, "The Leadership of the Reclamation Movement, 1875–1902" (PhD diss., University of California, Berkeley, 1951).

4. *New York Times*, November 30, 1902, March 9, 29, 1903; Mikel H. Williams, *The History of Development and Current Status of the Carey Act in Idaho* (Boise: Idaho Department of Water Resources, 1970), 69, 71; *Twin Falls News*, September 27, 1907; *Shoshone Journal*,

September 13, 1907, September 17, 1909.

5. Heber Q. Hale to Governor James H. Brady, August 23, 1910, James H. Brady Papers, Governors' Files, Idaho State Archives, Boise (hereafter ISA).

6. W. B. DaJarnatt to James H. Brady, April 16, 1909, Brady Papers; *Pittsburgh Directory of Directors* (Pittsburgh: R. L. Polk & Co.,1913), 203, 204; *The National Cyclopedia of American Biography* (Ann Arbor: University Microfilms, 1967 [1947]), 33:565–66; *Twin Falls News*, May 12, 1905, June 14, 1907; John C. Beebe, "Water Power and Its Relation to Irrigation in Southern Idaho," *Association of Engineering Societies Journal* 54 (February 1915), 69–72.

7. William Darrell Gertsch, "The Upper Snake River Project: A Historical Study of Reclamation and Regional Development, 1890–1930" (PhD diss., University of Washington, 1974), 70–71: *Seventh Biennial Report of the State Engineer to the Governor of Idaho* (Boise: Syms-York Co. 1908), 184–185; *Shoshone Journal*, March 8, 1907; *North Side News* (Jerome, ID), April 15, 1909.

8. Testimony of Kuhn's subordinates, J. H. Purdy, in "Plaintiffs Proposed Statement of Evidence," March 8, 1918, and G. L. Edwards, in Idaho Board of Land Commissioners Records, November 30, 1916 (copy), Twin Falls Salmon River Land and Water Company v. M. Alexander, et al., Idaho Attorney General Case File X–326, ISA; *New York Times*, July 8, 1913.

9. Fred A. Voigt to J. H. Brady, July 13, 1910, J. P. Whitla to State Board of Land Commissioners, July 22, 1910, Brady Papers.

10. In 1909, Brady summed up the board's impression of the Kuhn brothers: "they have never undertaken anything which has the least semblance of failure in the construction or operation of the same," Brady to W. B. DaJarnatt, April 30, 1909, Brady Papers.

11. *Seventh Biennial Report of the State Engineer*, 184.

12. Minutes of State Board of Land Commissioners, August 16, 1910, Brady Papers. Kuhn later relinquished several land areas that he judged unsuited to reclamation, so he ultimately attempted to reclaim 261,945 acres in the North Side project, Williams, *Carey Act*, 68.

13. Copies of contracts with Kuhn dated April 15 and August 15, 1907, Moses Alexander Papers, Governors' Files, ISA; James Stephenson Jr., *Irrigation in Idaho*, U.S. Department of Agriculture Bulletin 216 (Washington, DC: Government Printing Office, 1909), 44–45; *Irrigation [in] Twin Falls Country* (Twin Falls North Side Land and Water Co., 1909), 15, 18–19.

14. Archibald C. Milner to C. C. Baird, October 20, 1910. Baird to James P. Whitla, October 22, 1913, Twin Falls Land and Water Company Papers, Idaho State Historical Society, Boise (hereafter ISHS); "History of the Minidoka Project, Idaho to 1912" (typescript), 1:42, Project Histories and Reports of Reclamation Bureau Projects: Minidoka Project, Idaho, 1910–1919, Records of the Bureau of Reclamation, U.S. Department of the Interior, Record Group 115, National Archives, Microcopy 96, Roll 96 (hereafter MPHR and roll number); F. A. Banks, "Jackson Lake Storage," in *Proceedings of the Joint Conference of Irrigation, Engineering and Agricultural Societies of Idaho* (Twin Falls: Kingsbury Printing Co., [1919]), 108–109.

15. *Irrigation [in] Twin Falls Country*, 18.

16. Minutes of State Board of Land Commissioners, July 22, 1910, Brady Papers; *North Side News*, February 14, 1909. For Powell's ideas on this point, see Davison, "Leaders of the Reclamation Movement," 47.

17. *Chicago Record-Herald*, January 10, August 1 (special supplement on southern Idaho), August 13, 1909; *North Side News*, December 24, 1908; *Rupert Pioneer-Record*, November 3, 1909; S. H. Hays to F. A. Voigt, October 13, 1908, Voigt to A. L. Fox, August 12, 1909, Twin Falls Land and Water Company Papers.

18. *Shoshone Journal*, August 9, 1907; *Rupert Pioneer Record*, April 9, 1908; *Pocatello Tribune*, January 14, 1909; F. H. Newell to James H. Brady, January 22, 1909, Brady Papers.

19. F. C. Rogers to King Hill Irrigation and Power Co., August 31, 1908, King Hill Extension Irrigation Company Papers, ISHS.

20. Ern G. Eagleson to Charles Addison Beach, April 15, 1907, Eagleson to H. H. Bassler, October 7, 1907, Ernest G. Eagleson Papers, ISHS; *North Side News*, May 26, 1910; *Shoshone Journal*, April 2, 1909.

21. John Martin to C. H. Hammett, September 12, 1908, King Hill Extension Irrigation Company Papers; Fred R. Reed to James H. Hawley, September 4, 1911, S. J. Rich to James H. Hawley. November 25, 1911, A. E. Robinson to State Board of Land Commissioners, September 11, 1912, James H. Hawley Papers, Governors' Files, ISA; Mae Schultz to Moses Alexander, July 7, 1916, Alexander Papers; *Salt Lake Tribune*, October 6, 1907.

22. Frank S. Reid to W. H. Gibson, November 13, 1910, Brady Papers; S. D. Taylor to F. W. Buttschau, April 19, 1913, John M. Haines Papers, Governors' Files, ISA; *Seventh Biennial Report of the State Engineer*, 204; *North Side News*, March 4, April 8, June 3, 23, 1910; *Twin Falls News*, July 18, 1912.

23. *North Side News*, December 15, 1910; *Twin Falls News*, July 18, 1912; *Rupert Pioneer-Record*, November 28, 1912.

24. Idaho Irrigation Company, Ltd. v. Charles William Dill, March 7, 1914, 25 *Idaho Reports* 711–721; I. E. Bennett v. Twin Falls North Side Land and Water Company, Ltd., July 7, 1915, 27 *Idaho Reports* 643–655; Minutes of State Board of Land Commissioners, October 7, 1910, Brady Papers; S. D. Taylor to F. W. Buttschau, April 19, 1913, Haines Papers.

25. Annie Pike Greenwood, "Letters from a Sage-Brush Farm," *Atlantic Monthly* (September, 1919), 124:319; N. Jenness to F. O. Jillett, January 28, 1915, copy in William E. Borah Papers, Library of Congress, Washington, DC; F. W. Buttschau to John M. Haines, April 15, 1913, Haines Papers.

26. *Lincoln County Times* (Shoshone), December 27, 1911; L. T. Alexander to Edward B. Witwer, July 30, 1913, copy in Haines Papers.

27. Alexander to Witwer, July 30, 1918.

28. *North Side News*, November 17, December 1, 15, 1910; Minutes of State Board of Land Commissioners, August 10, 1910, Jesse Hawley to State Land Board, September 28, 1910, Brady Papers; L. T. Alexander to James H. Hawley, July 13, 1912, Hawley Papers; *Ninth Biennial Report of the State Engineer to the Governor of Idaho, 1911–1912* ([Boise, 1912]), 46–47.

29. *North Side News*, October 20, November 3, 1910; Joseph Collins to James H. Hawley, July 18, 1912, Hawley Papers.

30. Minutes of Board of Land Commissioners, September 10, 1909, July 22, 1910, Brady to R. A. Ballinger, February 26, 1910, Brady Papers; F. E. Weymouth and A. J. Wiley to F. H. Newell, March 1, 1910, MPHR, roll 101; Banks, "Jackson Lake Storage," 108.

31. *North Side News*, July 28, 1910.

32. *Ninth Biennial Report of the State Engineer*, 46, 173; "Report Prepared for Consulting

Board" (typescript), September 11, 1911, 7–8, A. P. Davis and others to F. H. Newell, February 25, 1911, MPHR, roll 101.

33. W. M. Wayman and others, "Report of the Engineering Commission" (typescript), September 3, 1912, Eagleson Papers.

34. "Report of the Idaho Irrigation and Drainage Code Commission of [on] the Twin Falls North Side Carey Act Project" (typescript, n.d.), Alexander Papers.

35. Minutes of State Board of Land Commissioners, April 27, 1910, Will H. Gibson to Land Commissioners, December 21, 1910, Brady Papers: A. E. Robinson to Land Commissioners, September 11, 1912, Hawley Papers; *Ninth Biennial Report of the State Engineer*, 47: *Wendell Irrigationist*, September 13, 1912. The board's fears were justified. On April 15, 1918, an Idaho district court apportioned Snake River waters to the North Side Project based on 1905 and 1910 priorities. In effect, several other projects were given better rights to the water than was the North Side Project.

36. A. P. Davis and F. E. Weymouth to F. H. Newell, February 14, 1913, Board of Engineers to Chief of Construction, May 24, 1915, MPHR, roll 101; D. R. King to State Board of Land Commissioners, September 15, 1914, Alexander Papers.

37. The Land Board approved this small quantity of water for 100,000 acres, about 1.5 acre-feet annually, on grounds that elsewhere in Idaho project farmers had subsisted on such amounts.

38. S. D. Taylor to F. W. Buttschau, April 19, 1918, Haines Papers.

39. *Pittsburgh Chronicle-Telegraph*, July 7, 8, 9, 1913; *New York Times*, July 8, 1913; Williams, Carey Act in Idaho, 73–74; "Explaining the Pittsburg[h] Crash," *Literary Digest* (July 19, 1913), 47:85.

40. *New York Times*, October 16, November 13, 1913, June 27, November 14, 17, 19, 1914; Idaho Board of Land Commissioners Records, November 30, 1915, copy in Idaho Attorney General Case File X-326, ISA.

41. Haines to F. B. Tichnor, August 1, 1914, Haines Papers.

42. Lane, "Our Paternal Uncle: A Study of Western Relations, How Uncle Sam Helps Those Who Help Themselves," *Sunset* (September, 1914), 33:518; F. H. Newell to S. D. Taylor, January 23, 1914, Lane to Haines, February 6, 13, 1914, A. P. Davis to Haines, December 4, 1913, Haines Papers.

43. Banks, "Jackson Lake Storage," 109; "Minidoka Project History, 1915" (typescript), 4, 80–81, MPHR, roll 98; "Report of Operations and Maintenance, 1914" (typescript), 20, MPHR, roll 102.

44. Minutes of State Board of Land Commissioners, January 29–February 3, 1916, Alexander Papers; W. G. Swendsen to D. W. Davis, May 9, 1919, David W. Davis Papers, Governors' Files, ISA; *Proceedings of the Joint Conference of Irrigation, Engineering and Agricultural Societies*, 47.

45. *Gem State Rural* (Caldwell) (February 1, 1917), 22:24; *Wendell Irrigationist*, September 25, 1919, October 28, 1920; *Seventh Biennial Report of the Department of Reclamation, State of Idaho, 1931–1932* ([Boise, 1932]) 35; *Jerome County Journal*, January 28, 1926.

46. *First Biennial Report of the Department of Reclamation, State of Idaho, 1919–1920* ([Boise, 1920]), 17; *Third Biennial Report of the Department of Reclamation, State of Idaho, 1923–1924* ([Boise, 1924]), 6; Williams, *Carey Act in Idaho*, 74.

47. Minutes of State Board of Land Commissioners, September 10, 1919, Davis Papers; "Minidoka Project History, 1919" (typescript), 12, 141–144, MPHR, roll 100; Gertsch, "Upper Snake River Project," 190–191.

48. In 1935, this proposal for federal control of the project was revived, but the Bureau of Reclamation discouraged the petitioners. George C. Sanford to Marshall W. Dana, March 4, 1935, Idaho Reclamation Records, ISA.

49. *Wendell Irrigationist*, April 16, 1920; R. E. Shepherd, "The Financing of Irrigation Development by Private Capital," *Transactions of the American Society of Civil Engineers* (June, 1927), 90:710–711; F. E. Schmitt, "Through the Reclamation Country," *Engineering News-Record* (November 16, 1923), 91:800.

50. Report of the Idaho Irrigation and Drainage Code Commission to the Governor of Idaho, 1915 (Boise, 1916), 73; F. E. Schmitt, "Through the Reclamation Country," *Engineering News-Record* (October 25, 1923), 91:677.

51. D. C. Henny et al., "Some Phases of Irrigation Finance," *Transactions of the American Society of Civil Engineers* (1928), 92:547.

CHAPTER 6

Idaho's White Elephant: The King Hill Tracts and the United States Reclamation Service

Progressive era reformers envisioned increased economic opportunity and social betterment for Americans when millions of acres of the West's arid land became irrigated farms under provisions of the National Reclamation Act of June 17, 1902. Many of these reformers interpreted the new statute, also known as the Newlands Act, as authorization for the United States Department of the Interior to undertake land reclamation work as speedily as possible by putting engineering, geological, and environmental principles ahead of politics. Frederick Newell, the first head of the U.S. Reclamation Service (USRS), an arm of the Interior Department, even boasted that the act pried "out of the hands of western politicians any excuse for interference" or pretext for "attempting to dictate in this work." But the same reformers claimed later—as did certain progressive era scholars and analysts of federal land policies and practices—that because science frequently took a backseat to politics after 1902, the USRS could not carry out unhindered its social and scientific missions during the next thirty years.[1]

Such a claim continues to exercise considerable appeal as an explanation of the Reclamation Service's shortcomings. However, plenty of evidence suggests that Newell and his subordinates never really discounted the political implications of their acts, nor did they actually try to remain disinterested technicians only to become hapless victims of greedy westerners and venal politicians. Furthermore, USRS leaders' political ineffectiveness and failure to anticipate changes in the American economy and society do not necessarily mean that their weaknesses stemmed from their self-proclaimed obsession with science and efficiency. In fact,

Man beside a spillway from an irrigation canal and gates. *Twin Falls Public Library, Clarence E. Bisbee Collection, 1153*

after 1910 the Interior Department reportedly retreated from its resolve to subordinate politics to science, a change sparked by several notorious incidents that arose from the failures of reclamation projects established under the Carey Act of August 18, 1894. Much historical evidence indicates that USRS failures were due in part to other factors. In responding to the difficulties besetting these ventures, the Reclamation Service often displayed its own ineptitude and became a victim of its own imperialist tendencies. Officials seized on the misfortunes of others in order to torpedo private enterprise reclamation, absorb projects that competed with or blocked federal efforts, and manipulate conditions to supplement the reclamation fund created by the 1902 irrigation law.

During the bad times plaguing many Carey Act projects, state leaders—particularly in Wyoming, Colorado, Idaho, and Oregon—inadvertently invited complications by demanding that federal reclamationists contribute at least financial resources toward resuscitating failing ventures. In Idaho, such pleas for federal intercession extended to projects of dubious merit and, in one instance, eventually led to a federal bailout of the bankrupt and badly engineered King Hill tracts. This case exemplified the economic pitfalls of arid land reclamation in the West.[2]

Two King Hill tracts in Elmore County, contiguous and stretching west to east for over twenty miles along the bottom of the Snake River canyon, emerged as Carey Act projects in 1904 and 1908. Private enterprisers originally anticipated reclamation of 17,666 acres within the first tract—developed by the Kings Hill Irrigation and Power Company—but later scaled down the venture to 13,462 acres. Virtually the same group of investors organized the Kings Hill Extension Irrigation Company in 1908 and promised water enough for 9,454 acres in this second project.[3]

Compared to most Carey Act projects in Idaho, the King Hill tracts were small in acreage, and they were uncommon in other ways as well. Most important, the irrigation plans required unusually complex engineering. Although the Snake River flowed alongside both tracts, architects decided against damming the river to create a reservoir to fill canals. A later attempt to use Snake River water by pumping it into canals proved financially unfeasible, due to limitations in the technology available. Instead, planners banked on diverting water from the Malad River through a canal stretching fifty-two miles from the river to the most distant point of the King Hill projects. Because of the uneven terrain, the Malad water had to flow twice through pipes suspended across the circuitous Snake and run uphill through siphons at several other points. For want of better sites, engineers had to dig parts of their canals into hillsides towering above the valley bottom and at other places, build their new waterways in flatter but still uneven terrain. Gullies were spanned as inexpensively as possible with wooden flumes, from which irrigation water leaked after a few seasons of use. To the further dismay of King Hill enterprisers and their engineers, all too often builders necessarily anchored flumes in canal banks, where the soil tended to become unstable when irrigation water permeated it.[4]

Such difficulties made the cost of water for King Hill farmers double that for settlers at other Carey Act projects in Idaho. For instance, at the Twin Falls South Side project, settlers paid $25

Victor Walker's homestead shack on the King Hill Project. *Idaho State Historical Society, 625-50.35*

per acre for water: King Hill farmers were charged $65 per acre, payable in ten annual installments.[5]

Promoters unblushingly quoted the high prices to prospective settlers, and they lured hundreds to the King Hill tracts by claiming that "any kind of fruit grown in the temperate zone will grow [here] to perfection." According to these boosters, perfection became possible at King Hill on account of fertile soil and the location deep within the Snake River canyon, where farms were protected from inclement conditions like frost that elsewhere in the state made producing crops such as berries and melons a risky undertaking. Forty acres of King Hill land "planted to orchard," promoters promised, would "produce in actual money returns more than [would] 320 acres of the best wheat and hay land found anywhere in the world."[6]

With lofty expectations, settlers occupied the land beginning in 1908, but all parties in these ventures learned during the next three years that wishful thinking and high water prices guaranteed neither good crops and irrigation services for farmers, nor riches for promoters and outside investors. The projects never lived up to expectations, partly due to afflictions common to Carey Act projects throughout the West, among them endemic land speculation and increasing reclamation costs. But the problems that had surfaced at King Hill in only three years tended to be primarily technological in nature. Project promoters had diverted insufficient amounts of Malad River water to satisfy farmers. Worse yet from agrarian viewpoints, the new irrigation system malfunctioned so often that farmers received no moisture at all during critical points in crop-growing seasons. Unpleasant surprises also greeted groups who pioneered at King Hill without first studying the area's geology. For instance, they discovered that a thin covering of topsoil concealed thick layers of sand beneath many sectors of the tracts, and much of the farmers' irrigation water percolated into these sand formations instead of sustaining crops.[7]

By the most conservative estimates, correcting such problems would require the water companies to rebuild parts of the irrigation system at a price of $188,000. State authorities were far more pessimistic and added half a million dollars the estimate, believing that reconstruction costs could total $688,000. Moreover, national economic conditions appeared to make it nearly impossible for the companies to raise any of this money and even put them at risk of bankruptcy. First, those investors whose $1.2 million had initially capitalized the King Hill tracts hesitated to risk more of their wealth on irrigation bond issues. They held their purses even more firmly after farmers rebelled against the inadequate water supply by refusing to pay for water in 1910 and 1911. This agrarian insurgency soon encouraged all buyers of western irrigation company paper to look even more askance at the King Hill tracts. Finally, the King Hill companies, already deeply in debt, attempted to pacify the farmers without improving the irrigation system, and at the same time to deflect the companies' other creditors. But these efforts placated neither agrarians nor investors. The Kings Hill Irrigation and Power Company became "hopelessly insolvent" in 1912, and a federal court placed it in the hands of a receiver on March 11, 1913. Bankruptcy overtook the Kings Hill Extension Irrigation Company the next year.[8]

In spite of these troubles, Idaho authorities like Governor John Haines (1913–14) speculated that little was actually lost at King Hill from the standpoints of the public weal and settler well-being. These officials counted on bondholders of the bankrupt corporations protecting their investments, and the only feasible way of doing so entailed new spending. According to such reasoning, bondholders had no other choice. Unless they bettered the irrigation system so that farmers would honor the water contracts, there was little chance that they would see any returns on their investments. In contrast to the governor and other officials, King Hill landholders entertained doubts about new investment, soon grew less sanguine about ever securing good irrigation works, and finally despaired of receiving dependable water supplies or recovering damages for lost crops. Many abandoned their farms, one of them protesting that conditions at King Hill provided only a "living on mountain air and scenery." As for the remaining settlers, an observer noted, they tended to become an "indifferent group" among whom successful farmers were "few and far between."[9]

Many observers shared the agrarians' distrust of new "high financiering" by bondholders to save the day. In fact, a journalist stated, reclamation of arid land at King Hill had devolved into "a license" for entrepreneurs "to exploit the public." Furthermore, this deploring of free enterprise seemed even more justified when, after nearly a year elapsed, bondholders still had pledged no money to pay for improving the irrigation system. In the same year, too, a federal district court judge ordered the Kings Hill Irrigation and Power Company to redeem all of its debts by March 10, 1914, but prospects for complying seemed so remote that liquidation of the company to satisfy its creditors had become inevitable. The bondholders had again kept their pocketbooks closed despite the near certainty of sacrificing their investments in the corporation's downfall.[10]

Critics continued to deprecate Governor Haines's reliance on private enterprisers and, at the last, cited the King Hill corporations' troubles as justification for federal intervention. They also saw no better way than federal intercession to salvage several other irrigation projects in Idaho. Those ventures, which were about as badly planned and ill financed as the King Hill tracts, had encountered new difficulties when the national market for irrigation company securities collapsed during 1911 and 1912.[11]

Calls for federal involvement were welcome tidings to Frederick Newell, who dreamed of acquiring larger realms in the West for the Reclamation Service. Accordingly, he encouraged the notion afoot that resolving the newest problems with Carey Act projects probably required federal aid. As for Idaho, Newell reportedly remarked during discussions with Boise journalists on October 20, 1913, and in statements before the Boise Commercial Club two days later that progress on the state's Carey Act projects had never been satisfactory, except at the Twin Fall South Side tract. Hence, he proposed a federal-state partnership in which the two governments pooled equal sums of public money to be expended on bettering the irrigation works for certain tracts, like those at King Hill, where the problems had become overwhelming. Newell defended his plan as workable and, for proof, alluded to results from certain equal-partnership agreements between the United States and Oregon governments. Already, Newell noted, $900,000 had been placed at the disposal of Oregon public officials under these contractual agreements, and they expected with this money to salvage one of the state's Carey Act projects.[12]

Given the King Hill tracts' low estate, a condition that even the projects' best defenders acknowledged publicly, and in light of troubles bedeviling other irrigation projects in Idaho, Newell expected his federal aid proposals to win support throughout the state. However, he misjudged the impact of his statements. His critics in Idaho denounced him for indicting so sweepingly free enterprise reclamation under the Carey Act. In particular, they fomented indignation against Newell for saying that only a single Idaho tract (out of over forty Carey Act projects authorized in the state at the time) was successful. When such outcries continued, Newell sought shelter from the storm with claims that journalists had misinterpreted his statements. Worse yet from his standpoint, many of his Idaho critics persuaded others to deny his proposals serious consideration, partly on grounds that he was out of favor with President Woodrow Wilson's administration. Such speculation rested heavily on the fact that Newell's power, even within his own bureau, had waned after Franklin Lane became secretary of the interior in 1913. Many Idahoans mistakenly presumed that the decline occurred because Newell's ideas about handling Carey Act reclamation problems ran counter to Lane's own thinking.[13]

At the same time, Newell's critics pointed out that his United States-Idaho partnership plan counted for nothing unless Congress and state legislators funded it. His proposal appeared especially unrealistic because Idaho's state constitution disallowed extending "the credit of the state" to help an "individual, association, municipality, or corporation." Governor Haines, aware of this clause, grudgingly reviewed plans for fulfilling Idaho's obligation under a federal-state compact by raising money from marketing state-issued bonds. He subsequently abandoned this bond scheme when he detected little support for a constitutional amendment to authorize it. Newell then retracted his offer. Embarrassed by rebuffs that he never anticipated, Newell denied that he ever extended a firm offer of federal assistance to Idaho. However, he still expected Lane to "look with favor on any proposition which might bring about a joining of state and federal forces in the better development of the West." He claimed that his controversial remarks in October 1913 at Boise were intended solely "to aid in laying the foundation for a better understanding of the difficult matter" of the collapsing Carey Act projects.[14]

Haines also discounted Newell's ideas for a federal-state compact because he continued to hope that federal intercession might be avoided. Although observers labeled the tracts a failure, Haines considered it reassuring that King Hill farmers actually irrigated about 5,000 acres during 1913. More important, he respected the judgment of S. D. Taylor, Idaho's Carey Act land commissioner, who argued that bondholders had every reason to beat a court-mandated deadline of March 10, 1914, for saving the Kings Hill Irrigation and Power Company from foreclosure and to refinance both "Kings Hill propositions" in order to preserve "their [original] investments."[15]

In the end, Haines relied on these bond holders' self-interest to bring the King Hill projects to full fruition. At the same time, he proposed to Lane that the federal government unilaterally complete the West End Twin Falls project. At this Carey Act tract, the irrigation system was unfinished, settlers had received practically no water for their crops, and the project's promoters despaired of selling enough corporation bonds to capitalize completion of the system.[16]

Within just a few weeks, it became clear that Haines erred in trusting capitalistic self-interest to save the Kings Hill Irrigation and Power Company and, in turn, ensure good times for King Hill landholders. Instead, the company's bond holders declined to open their wallets. After that refusal, other financiers declined even to bid a judicially mandated minimum of $30,000 for the property rights at a receiver's sale on March 10. This dearth of private bidders led Idaho state authorities to intervene at the last moment to ensure that the project would not be abandoned. They purchased the property rights at the lowest permissible price on grounds that good public policy required them to prevent King Hill settlers from losing all of their holdings. Observers generally judged the state's new property a white elephant, the acquisition of which Haines and his successor attempted to justify during the next few years.[17]

While answering critics who nagged them about acquiring undesirable assets, state leaders also wrestled with complications at King Hill. Operating the irrigation system proved progressively more difficult because the structures were "old and rickety" and "constantly needing repairs," according to Benjamin Shawhan, the state administrator of the tracts. Moreover, state spending at King Hill increased geometrically despite several governors' efforts to hold it down by utilizing the free labor of penitentiary inmates for maintenance work. King Hill cost Idaho slightly more than $82,000 during the state's first two years of stewardship.[18]

Officials fretted that the state might be forced into expending hundreds of thousands of dollars to rebuild and refurbish the system. Making the state-controlled project "first class," the state engineer estimated, entailed spending $300,000 just to upgrade twenty-seven miles of canal structures. George Archibald, Carey Act inspector for the United States General Land Office, calculated that administering the state's new property and putting the irrigation system in "good operating condition" would require Idaho to spend at least $320,000. Also, Archibald judged the state "morally responsible" for the Kings Hill Extension Irrigation Company project when it went on the auction block without much chance of attracting private purchasers. Such responsibility could become especially difficult to avoid because the extension project's land relied largely on Malad River water transported to the tract in the same canal that served the state-controlled project. State officials worried that total expenses for the two projects could soar past $400,000, and a panel of Reclamation Service engineers confirmed their misgivings. According to the federal engineers, the price could reach $825,000.[19]

Haines decided nonetheless that sound public policy required government aid to salvage both King Hill tracts and nurse them to prosperity. Faced with the high costs of keeping the tracts in business and the near certainty that the Idaho legislature would never underwrite such expenditures, he appealed to the secretary of the interior. First, Haines proposed that the Reclamation Service assume responsibility for improving the King Hill irrigation system and in certain instances for providing necessary new structures. The governor justified his proposal on the grounds that federal "cooperation" in such matters had become "absolutely essential" because "the day of irrigation development on a large scale by private or corporate enterprise [was] a thing of the past." He called, moreover, for "a very large appropriation" from Congress with which to complete western irrigation projects, among them Carey Act undertakings like the King Hill tracts, "that have not been financially successful."[20]

Reclamation Service officials welcomed Haines's call for a federal bailout of King Hill, but the secretary of the interior proved less receptive. Lane adverted to Newell's old proposals for a federal-state partnership, and he continued for several years to advocate such arrangements. He persisted primarily because of the low estate of the federal reclamation fund, an account that could never meet all demands on it. Lane considered the infusion of state money the best way to augment the federal coffers.

Meanwhile, Lane judged the most nettlesome projects to be situated in Idaho, Oregon, and Colorado. But before he allowed federal participation, he insisted that the prospective partners develop "sensible, practicable plan[s]" to deal with particular tracts that could become "meritorious" projects in the long run. Lane assigned this sorting of potentially successful from unlikely projects to an advisory group, the Interstate Irrigation Commission, which he organized in December 1914. His reluctance to take action continued until January 15, 1915, when under renewed pressure from Idaho's congressional delegation, he counter-proposed that the United States and Idaho governments reconstruct the King Hill irrigation system under conditions of "sharing equally" in all "responsibility." This response from Lane, according to Senator William E. Borah, indicated that Idaho authorities might never negotiate a federal take-over at King Hill. More favorable terms appeared unlikely because Lane had President Wilson's support in restricting new federal commitments to western reclamation.[21]

Dealing with Lane's counterproposal fell to Moses Alexander, who had just defeated Haines in the gubernatorial election of 1914, and to several others who composed the State Board of Land Commissioners (commonly known as the Land Board). This body served as public overseer of Idaho's Carey Act projects.

Although Alexander was a Democrat and Republicans numerically dominated the Land Board, all of them desired no less than federal hegemony at King Hill. To encourage this outcome, they were willing as a last resort to transfer Idaho's equity to the federal government without receiving compensation for the property rights. Consequently, the board rejected Lane's proposal to share responsibility. When Lane persisted, the board discontinued negotiations. These acts angered King Hill landholders, one of whom charged Alexander with believing that the projects were "defunct, hopeless and should be abandoned." The state's project manager at King Hill defended the board. He warned that Lane's plan would open the door to similar demands from other settlers on Carey Act projects and bankrupt the state.[22]

Such explanation mollified King Hill landholders scarcely at all, but their resentment ebbed when, later in 1915, the state legislature showed them—by authorizing the Land Board to transfer the projects to federal stewardship—that the state's fight for a federal take-over at King Hill had not ended. The lawmakers also approved a memorial to Congress asking for appropriations to finish the projects satisfactorily; this proposal envisioned landholders reimbursing the government for its expenditures at King Hill in the same ways that the federal government expected to recoup its outlays from residents of its more successful Payette-Boise and Minidoka projects.[23]

The case for federal intervention at King Hill grew even stronger during 1915 in the wake of an Idaho House of Representatives subcommittee's hearings on difficulties at Carey Act projects scattered across the Snake River Plain. At the public proceedings, witnesses excoriated Carey Act reclamation methods; these attacks received widespread notice, and the state legislature responded to the public outrage.

Most significant, legislators created the Idaho Irrigation and Drainage Code Commission, the principal mandates of which included finding ways to bring relief to settlers on Carey Act lands, and the commission examined the King Hill tracts early in its deliberations. The commission pointed out in its formal report on King Hill that optimism among project landholders had sagged after momentarily rising earlier in the year. Many were again "extremely discouraged," it reported, but they unanimously preferred federal take-over of the projects to other alternatives. The commission likewise concluded that federal stewardship would better serve the tracts. It urged state officials to resume negotiations with federal authorities but reviewed several alternatives to federal intercession. As a last resort, the commission stated, the state should "take up the projects," build adequate irrigation works, and compel reimbursement for such expenditures from the landholders.[24]

Governor Alexander agreed wholeheartedly with the commission's views about federal stewardship at King Hill, but dismissed its recommendation that the state rescue the projects if all else failed. He shuddered, as did other Land Board members, at the likelihood of similar demands flooding in from many other irrigators. But Alexander could make no headway persuading Lane to assume all responsibility for the King Hill tracts. Consequently, in 1916, Idaho's congressmen attempted to sidestep the secretary of the interior. They introduced legislation authorizing the federal government to place adequate irrigation works on the tracts. However, this bill languished in the House of Representatives. According to Senator Borah, its prospects remained dim because eastern voters objected to further federal spending on western reclamation ventures until settlers had "paid back" U.S. investments in the earlier projects.[25]

When congressional inaction on the King Hill bill persisted, Alexander reasoned that the legislation might be passed if its opponents became convinced that federal investments would be safe.

Consequently, Alexander, the Land Board, and its agents at King Hill pressured residents of the tracts to cooperate with a new stratagem; primarily, they expected landholders to establish a 24,000-acre irrigation district. Idaho laws authorized irrigators to establish by majority vote among themselves their own district to build irrigation works and distribute water. If the proposed district were eventually enlarged to include virtually all of the acres King Hill visionaries had ever eyed for reclamation, theoretically it could easily collect enough money from landholders for the next twenty years—and where necessary flex the district's statutory taxing powers over them—to reimburse the United States. King Hill electors acquiesced by a vote of 479 to 48 on February 27, 1917, and the new irrigation district opened for business within a few weeks.[26]

Meanwhile, Alexander continued his attempts to counter Lane's opposition. In 1916, Lane agreed to authorize new studies of the King Hill tracts by his own Reclamation Service engineers. Alexander also procured for Lane's edification a study of the tracts done by private consulting engineers. Both groups essentially upheld the 1914 assessment made by the Carey Act inspector George Archibald that only governments could salvage the projects. Then, armed with the reports, Alexander appealed to Arthur Powell Davis, successor to Newell at the USRS and one of the federal reclamationists to whom Lane had increasingly listened after 1914. Alexander counted on the new engineering studies so impressing Davis that he would recommend federal stewardship at King Hill and persuade Lane to make provisions for it in his next budget request to Congress.[27]

Davis expressed interest in the new assessments of King Hill, but in the end Alexander's maneuvering fell short of its mark. Lane, who still believed as strongly as ever that the fiscal straits of the federal reclamation fund inevitably compelled states to share the monetary burdens of dealing with the West's troublesome Carey Act projects, reiterated his old preferences for federal-state partnerships. He also added new objections to handling these problems any differently. He stated that King Hill initiatives in Congress had raised new policy questions for him to ponder; he went on to say that before supporting any King Hill legislation, he would first have to determine how far he might properly go since Congress had recently deprived him of his old discretion to allocate money from the fund.[28]

The governor concluded that the state must present its case more aggressively in Congress, but he judged Lane's new obstructionism as proof that the Interior Department was his greatest hindrance. Alexander also decided that dealing with this problem required better tools. Happily for him, the outcome of the 1916 presidential and state elections appeared to have just provided such aids. Of particular importance to his designs, the elections brought fundamental political realignments in several far-western states. In Idaho, Democrats bettered their position even more significantly than in many other states. Woodrow Wilson captured 52 percent of the popular vote there. Alexander won reelection in 1916, although not by a majority that permitted him much crowing, and for the first time in more than a decade, Democrats controlled the state legislature.[29]

Making the most of these developments, Alexander reminded Lane that Idaho had changed from a Republican state to a Democratic state, that Wilson had carried Idaho handily, and that Idaho Democrats "now want your help to make our [campaign] promises good." For starters, the governor demanded Lane's "full influence toward having the King Hill Irrigation Project taken over by the government." Moreover, Democrats controlled the Idaho Land Board after 1916, and they commissioned James Hawley (who as governor from 1911 to 1912 was one of only four Democrats to hold that office between 1890 and 1919) to lobby at the national capital. Alexander also converted the King Hill issue into a crusade by insisting that Idaho Republicans who sat in Congress, notably Borah and Congressman Addison Smith, put politics aside and cooperate with him.[30]

Lane rebuffed Alexander at first, but soon he and a number of Wilsonian congressmen relented in the face of the governor's persistence, his claim to a share of the Democratic Party's political spoils, and logrolling by Idaho's Republican senators and representatives. In 1917, Congress accepted the USRS projection that spending $1 million would cure King Hill's irrigation woes, then appropriated the first $600,000. Early in the dealings, Alexander also hoped to recover for the state $100,000 that it had expended at King Hill after 1914, but he backed off in order to secure what he most wanted. In return for the federal spending and $1 (to make the transfer a legal sale), Idaho relinquished by quit-claim deed its suzerainty and property interests at King Hill.[31]

Reclamation Service leaders welcomed the new addition to their domain, for the old expansionist impulses within the bureau had not at all diminished following Newell's departure. USRS officials, state authorities, and irrigation district officers negotiated contracts spelling out conditions for eventually spending $1 million on irrigation infrastructures in exchange for landholder repayment of a like amount in twenty annual installments. The agreements required the irrigation district to guarantee repayment, and because the district held statutory power to tax all property owners residing within its boundaries, full reimbursement appeared to be ensured. Operational control of the irrigation works went from the state of Idaho to the irrigation district. Alexander and his Land Board colleagues insisted that this divestiture happen speedily. It became effective about a year later, on Apri1 1, 1918.[32]

A journalist described these arrangements as obliging the Reclamation Service to overcome "melancholy evidences of past failures" that "strew the hillsides and canyon slopes" at King Hill "in the form of decayed wooden flumes, old pipes, and ruined bridges." Nonetheless, landholders anticipated the imminent arrival of good times, speculators assumed that prosperity lurked around the corner, and farmland prices rose to $200 per acre in 1920. Also, to Alexander, the Idaho Land Board, and Reclamation Service administrators, King Hill's future seemed far brighter than a handful of doomsayers predicted. These architects of a new day assumed that $1 million would upgrade the irrigation system, making a 16,000-acre project technologically feasible and, in turn, large enough to accommodate sufficient numbers of farms so that each settler's share of the debt would be reasonable. The architects predicated such calculations on several assumptions: that the federal role at King Hill centered on rebuilding the common canal serving both tracts, bringing its carrying capacity to 300 second-feet; that the increased flow would meet the needs of both tracts; and that, further to restrict financial burdens on settlers, the irrigation district would refurbish the remainder of the system and keep canals and laterals in operable condition less expensively than could the federal govemment.[33]

Although nearly all King Hill landholders trusted federal reclamationists' methods at the outset, their optimism lasted scarcely two years. Starting in 1919, farmers entertained misgivings, and their doubts changed to dismay as it became evident that USRS officials had underestimated the hurdles. One engineer reported that, after his bureau spent the first $600,000 out of the $1 million allocation, the main canal embankment still contained many "weak and dangerous" sections awaiting repairs. He added that wood siphons and flumes needed replacing and that laterals branching from canals remained generally in a "very dilapidated condition." In short, additional costly renovation was necessary.[34]

The Reclamation Service (reorganized in 1923 as the Bureau of Reclamation) expended nearly $1.4 million between 1918 and 1923 eliminating such defects. Furthermore, costs continued to soar until federal outlays became twice what state and federal architects had prescribed in their grand salvation plans of 1917. In part, this doubling of federal spending reflected the larger role the Reclamation Service assumed when the irrigation district failed to fulfill some of its obligations.[35]

Admiration of federal reclamation methods at King Hill also subsided after 1920 on counts other than just the miscalculation of the extent and cost of bettering the irrigation system. For one, critics accused bureau scientists of perpetuating fundamental flaws from the 1917 blueprints. At that time, planners forecast irrigation of 16,000 acres, a figure that rested primarily on economic rationales, and Reclamation Service technocrats subsequently accepted this goal and strove conscientiously to reach it. But critics contended that the tracts lacked sufficient water with which to irrigate so large an acreage. According to some King Hill landholders, irrigating 16,000 acres required at least 500 second-feet. They believed that federal scientists should have detected such mistakes in the original blueprints and, instead of adhering to those plans, redesigned the irrigation system in order to increase substantially its capacity. Irrigation district leaders demanded that federal reclamationists deliver "an adequate supply of water" before settlers actually reimbursed the government for any of its spending at King Hill.[36]

Reclamation Service administrators considered such criticism unwarranted, especially after the bureau had actually improved canals, built laterals, installed new siphons and steel flumes, and placed a concrete lining in miles of waterways. The new lining, federal reclamationists pointed out, had already ensured that mishaps like canal and lateral embankments collapsing were no longer endemic. Elwood Mead, U.S. reclamation commissioner, argued that the King Hill tracts should have prospered. However, this was not the case. Population growth declined to a standstill after 1920, the lots of agrarians worsened, and observers described tellingly how local poverty became increasingly manifest as the agricultural depression of the twenties also exacted its toll. Farmers began to desert their land in such numbers that only 7,000 of 16,000 acres stayed in cultivation. As of 1922, according to the federal reclamationists, the King Hill tracts contained 175 farms on which 599 people resided. The four towns within the projects' boundaries had a total population of 2,052.[37]

Viewed from afar, it appeared that King Hill agrarians had prospered with other American farmers when land and crop prices rose during World War I in response to demands for food in Europe. All now suffered from the collapse of prices following the end of the war. Nonetheless, such comparisons could not console settlers who faced paying the consequences of others misjudging reclamation costs.

When landholders abandoned the King Hill tracts in droves after 1920, nobody replaced them, and the remaining farmers could never reasonably pay the federal piper. As of June 30, 1923, the holders of titles and Carey Act claims to 16,000 acres owed the United States $2 million. But, an agricultural economist calculated, compelling full restitution from those who could pay meant increasing those landholders' payments from $137 to $292 per acre.[38]

Such prices would place this water generally beyond the reach of King Hill farmers, even after crop prices improved as the agricultural depression subsided. Water costs of such magnitude also seemed preposterous when other Idaho farmers, even those in troubled projects, paid a fraction of that amount. For instance, at the Boise site, which was likewise a federal undertaking, bitter battles between settlers and the Reclamation Service had just ended in 1921 with settlers grudgingly accepting charges of about $77 per acre.[39]

Conditions for King Hill landholders improved hardly at all despite better times for farmers nationally after 1924, and disputing between agrarian debtors and their federal creditor intensified. Finally, a special advisory group to the secretary of the interior applied to King Hill the Fact-Finders Act of December 5, 1924. The new federal law based reimbursement for reclamation costs on settlers' anticipated annual returns, and by such standards, the special advisory group wiped out half of the King Hill landholders' $2 million indebtedness, inasmuch as their prospects for crop production had declined so precipitately. A Board of Survey and Adjustment, created by Congress to review the advisory group's decisions, further whittled down the balance to $531,938. However, the board also projected future losses of up to $287,024, largely because the amounts of water available at King Hill met irrigation requirements for no more than 12,500 acres and likely for only 10,000 acres over the long haul. Coming next in this process of arranging final settlements, Congress wrote off all King Hill indebtedness in 1926 and, in 1934, transferred federal property rights to the irrigation district.[40]

Carey Act enterprisers, King Hill farmers, Idaho public authorities, and Reclamation Service technocrats learned, in turn, that none of them stood much chance of creating an economically viable project. Several factors—the topography of the land, the tricky geology of the region, and the Snake River's meandering course through a narrow valley bottom—made transporting Malad River water a highly complicated and extremely expensive undertaking. However, such conditions existed at other western project sites and should have served as warning for King Hill reclamationists.

Real winners never emerged at King Hill, but the actions of Governor Alexander and the state Land Board ensured that Idaho incurred the least possible debt. By transferring their stewardship obligations to federal hands, Idaho authorities largely sidestepped the economic con sequences. Not without justification, an observer likened their ridding the state of its white elephant at King Hill to "a man trying to let go of a bear's tail."[41]

For its part, the United States Reclamation Service enthusiastically adopted goals at King Hill that were not, for all the bureau's scientific expertise and engineering acumen, readily achieved at sensible costs. Taxpayers paid dearly when their elected officials permitted Idaho to transfer responsibility for King Hill to the federal government. But, in the end, the bureau's sins consisted of more than grasping greedily for a larger empire everywhere—to the point of taking in others' white elephants. Its officials erred in drawing irrigation blueprints without first carefully assessing problems common to reclamation projects during the early twentieth century. Were it better forewarned along these lines, the bureau presumably could have been better forearmed in tackling reclamation problems at King Hill.

This article originally appeared in *Pacific Northwest Quarterly* 83, no. 1 (January 1992). Reprinted with permission.

Notes

1. F. H. Newell to D. W. Ross, September 28, 1903 (qtns.), file 232, box 2/SB 203133, Acc. No. 63-B-0157, Bureau of Reclamation Records, RG 115, Federal Archives and Records Center at Denver; James Penick Jr., *Progressive Politics and Conservation: The Ballinger-Pinchot Affair* (Chicago: University of Chicago Press, 1968), 63–66; Gene M. Gressley, "Arthur Powell Davis, Reclamation, and the West," *Agricultural History* 42 (1968), 241–42.

2. The Carey Act of August 18, 1894, authorized ten western states and territories each to withdraw one million acres of potentially irrigable land from the public domain and arrange for reclamation of such land at the hands of private enterprise. States authorized entrepreneurs to build irrigation works and sell water to settlers at a profit. In Idaho, entrepreneurs eventually proposed reclamation of about three million acres. For accounts of such projects, see Mikel H. Williams, *The History of Development and Current Status of the Carey Act in Idaho* (Boise: Idaho Department of Reclamation, 1970), and Hugh T. Lovin, "The Carey Act in Idaho, 1895–1925: An Experiment In Free Enterprise Reclamation," *Pacific Northwest Quarterly* 78 (1987), 122–33. [Chapter 4 in this book.]

3. [G. B. Archibald], "General Report on Kings Hill Project and Kings Hill Extension Project in Idaho" [1914?], 148,153, in "Project Histories and Reports of Reclamation Bureau

Projects: King Hill Project, Idaho, 1911–1926," M96, roll 61, RG 115, National Archives (hereafter cited as "Project Histories" with appropriate roll number); Williams, 47–48; Olive Groefsema, *Elmore County: Its Historical Gleanings* (Caldwell, ID: Caxton Printers, 1949), 368–69. At first, the companies involved in these projects used in their corporate names the words *Kings Hill*, a designation that referred to a prominent geographical feature of the area. This hill was always identified in Idaho as *King Hill* so far as is known, and it became commonplace to drop the s in public references to the corporations. For the fullest explanation of the origin of the name, see Federal Writers' Project of the Works Progress Administration (compilers), *The Idaho Encyclopedia* (Caldwell, ID: Caxton Printers, 1938).

4. Groefsema, 369–71; James Stephenson Jr., *Irrigation in Idaho*, U.S. Dept. of Agriculture, Office of Experiment Stations Bulletin 216 (Washington, DC, 1909), 51; U.S. Dept. of the Interior, U.S. Reclamation Service, "King Hill Project Idaho," Map No.1777 (1918), copy in James R. Garfield Papers, Library of Congress; "General Report on Kings Hill Project," 42–43,106–107, 175.

5. *Seventh Biennial Report of the State Engineer to the Governor of Idaho, 1907–1908* (Boise, 1908), 192 (hereafter cited as *Report of the State Engineer* with appropriate years). Within a few years, promoters of the King Hill tracts raised about $1.2 million on the strength of the settlers paying such amounts of money for their water: see "Statement Showing Amount [of] Securities of the Kings Hill Irrigation and Power Company, and the Kings Hill Extension Irrigation Company Held by the Farwell Trust Company" (n.d.), Kings Hill Extension Irrigation Company Papers, Idaho State Historical Society (ISHS), Boise. In the beginning, King Hill promoters estimated even higher water prices (Dorothy Lampen, *Economic and Social Aspects of Federal Reclamation*, Johns Hopkins University Studies in Historical and Political Science 48, no.1 [Baltimore: Johns Hopkins Press, 1930], 67).

6. A. M. [Alexander McPherson] to J. M. Whittaker, November 9, 1909 (1st qtn.), and Kings Hill Irrigation and Power Company to Rodgers and Rogers, September 3, 1908 (2d qtn.), Kings Hill Extension Papers; *Twin Falls (Id) News*, August 21, 1908.

7. Will H. Gibson to State Board of Land Commissioners, August 22, 1911, James H. Hawley Papers, Governor's Files, Idaho State Archives (ISA), Boise; W. G. Hoyt, *Water Utilization in the Snake River Basin*, U.S. Geological Survey, Water Supply Paper 657 (Washington, DC: U.S. Government Printing Office, 1935), 157.

8. *Report of the State Engineer, 1911–12*, 49; S. D. Taylor, *Carey Act Projects: Report on the Industrial Development of Idaho, Accompanied by the Reclamation of Desert Lands by Actual Irrigation under the Carey Act* (Boise, 1913), 13; "General Report on Kings Hill Project," 33–36; G. J. and C. C. Magenheimer to E. W. Kincheloe, November 9, 1912 (qtn.), Kings Hill Extension Papers.

9. Taylor, 13–14; *Report of the Idaho Irrigation and Drainage Code Commission to the Governor of Idaho, 1915* (Boise, 1915), 45; "Report of State Land Board Meetings," November 8, 9, 11, 1911 (1st qtn.), Hawley Papers; "Annual Project History [for 1921]," 149 (2d, last qtns.), in "Project Histories," roll 59.

10. *Caldwell (ID) Tribune*, August 29, 1913 (qtns.); "General Report on Kings Hill Project," 36–37.

11. By 1913, an Interior Department study concluded that "a general lack of confidence on account of past failures" had resulted in Carey Act projects' "inability to secure [additional] financial backing" (report on Carey Act projects in 62d Cong., 3d Sess., 1913, S.D. 1097, p. 22 [Serial 6365]). See also "Federal Report on Carey Act Irrigation Projects," *Engineering News* 69 (1913), 1321: and Ray Palmer Teele, *Irrigation in the United States: A Discussion of Its Legal, Economic and Financial Aspects* (New York: D. Appleton and Company, 1915), 133–34.

12. William Darrell Gertsch, "The Upper Snake River Project: A Historical Study of Reclamation and Regional Development, 1890–1930," (PhD diss., University of Washington, 1974), 89; *Idaho Statesman* (Boise), October 20, 23, 1913; *Twin Falls News*, October 23, 1913. For an account of efforts to salvage an Oregon reclamation project—an attempt that Newell believed Idaho authorities should emulate—see Martin T. Winch, "Tumalo—Thirsty Land, Part Three," *Oregon Historical Quarterly* 86 (1985), 153–82.

13. For accounts of Lane curbing Newell's power in the U.S. Reclamation Service, see Donald Worster, *Rivers of Empire: Water, Aridity, and the Growth of the West* (New York: Pantheon Books, 1985), 177; Samuel P. Hays, *Conservation and the Gospel of Efficiency: The Progressive Conservation Movement, 1890–1920* (Cambridge, MA: Harvard University Press, 1959), 248; Elmo R. Richardson, *The Politics of Conservation: Crusades and Controversies, 1897–1913*, University of California Publications in History 70 (Berkeley: University of California Press, 1962), 153.

14. Idaho Constitution, Art. 8, Sec. 2, copy in *Idaho Code Containing the General Laws of Idaho Annotated, Vol. 1* (Indianapolis, 1980), 249; F. P. King to C. P. Stephens, November 4, 1913, Idaho Reclamation Records, Collection AR-20, ISA; John M. Haines to J. E. Clinton, November 11, 1913, and F. H. Newell to S.D. Taylor, January 23, 1914 (last qtns.), John M. Haines Papers, Governor's Files, ISA.

15. Taylor to King, February 16, 1914, Haines Papers: Taylor, 13, 14 (qtns.). If the debts of the bankrupt Kings Hill Irrigation and Power Company were not redeemed before March 10, 1914, the company would forfeit its investments at King Hill, all of its debts would be canceled, and all of its properties at King Hill would be transferred to any bidder who purchased receiver's certificates worth $30,000.

16. Haines to Franklin K. Lane, January 15, 1914, Haines Papers. For accounts of the trouble-plagued West End project, see Frank K. Welles, *The Story of a Carey Act Investment* (Salem, OR: 1914); "History of the Carey Act Project Originally Known as the West End Twin Falls Irrigation Company's Project, Now Known as the Idaho Farm Development Company's Project," September 10, 1919, David W. Davis Papers, Governor's Files, ISA.

17. *Glenns Ferry (Id) Gazette*, March 6, 13, 1914; Gertsch, 90. Unlike Haines, state legislators reposed so little trust in the bondholders that, in 1913, they authorized the purchase with money from the state's Carey Act trust fund. State officials later defended the purchase partly on grounds of the need "to protect the good name of the state" after promoters had intimated in their advertising that Idaho guaranteed settlers against losses on Carey Act projects (*Gazette*, January 29, 1915). Kings Hill Irrigation and Power Company bondholders subsequently complained that they received only $9,000 from $30,000 realized "from the sale of the project" (E. T. Meredith to Haines, June 20, 1914. Haines Papers).

18. Benjamin Shawhan to Moses Alexander, June 20, 1916 (qtns.), and James Munro to Alexander, February 27,1917, Moses Alexander Papers, Governor's Files, ISA.

19. "General Report on Kings Hill Project," 296 (1st qtn.), 181 (2d qtn.), 187 (3d qtn.): D. C. Hanny, James Muon, and J. H. Miner to board of engineers, October 29,1917, in "Project Histories," roll 58; Shawhan to Alexander, January 16, 1915, Alexander Papers; *Gazette*, January 22, 1915.

20. J. H. Peterson to William E. Borah, March 21, 1914, William E. Borah Papers, Library of Congress; *Gazette*, October 30, December 11, 1914; Haines to F. B. Tichnor, August 1, 1914 (qtns.), Newell to Haines, November 21, 1914, Haines to Newell, November 28, 1914, and to Lane, December 17, 1914, all in Haines Papers.

21. Lane to Haines, February 13 (1st qtn.), July 13, 1914, Haines Papers; Franklin K. Lane, "Our Paternal Uncle: A Study of Western Relations, How Uncle Sam Helps Those Who

Help Themselves," *Sunset* 33 (1914), 512–18 (2d qtn., 518); [Benjamin P. Shawhan], "In Regard to King Hill" (n.d.), n. pag. (3d qtn.) and Borah to Orville C. Sanborn, February 26, 1915, Borah Papers.

22. Edward M. Clark to Borah, April 29, 1915 (qtn.); "In Regard to King Hill," and Shawhan to King Hill landowners, November 23, 1916, all in Borah Papers. Alexander to R. M. McCracken, February 4, 1917, Alexander Papers.

23. General Laws of the State of Idaho Passed at the Thirteenth Session of the State Legislature (Boise, 1915), Chap. 26 (S.B. 109), 79: Shawhan to Alexander, August 12, 1915, Alexander Papers; *Gazette*, April 2, 1915.

24. Report of the Idaho Irrigation and Drainage Code Commission, 44–48 (qtns. 47, 48); *Gazette*, October 22, 1915.

25. Sanborn to Borah, April 11, 1916 (copy), and Borah to P. E. Desault, December 1, 1916, Borah Papers.

26. Shawhan to King Hill landowners, November 23, 1916; *Gazette*, March 2, 1917; "History: King Hill Project, Idaho" (1917), in "Project Histories," roll 56.

27. Shawhan to Alexander, June 26, 1916, Alexander Papers. Shawhan to Addison T. Smith. June 15. 1916, and to Borah, August 3, 1916, and "In Regard to King Hill," all in Borah Papers.

28. Lane to Alexander, December 7, 1916, Alexander Papers; Shawhan to State Board of Land Commissioners (December 1916), Idaho Reclamation Records.

29. David Sarasohn, "The Election of 1916: Realigning the Rockies," *Western Historical Quarterly* 11 (1980), 285–305; Sarasohn, *The Party of Reform: Democrats in the Progressive Era* (Jackson, MS: University Press of Mississippi, 1989), 220.24; *Idaho Statesman*, November 11, 28, 1916.

30. Alexander to Lane, November 30, 1916, Alexander Papers; Alexander to Borah, December 1, 1916, Borah Papers; Minutes of the State Board of Land Commissioners, April 20, 1917, 3, Idaho Reclamation Records.

31. Lane to Alexander, December 7, 1916, Alexander to R. M. McCracken, February 4, 1917, and Smith to Alexander, February 26, March 6, April 4, 1917, Alexander Papers; *Gazette*, January 6, April 13, 1917; Williams, 50; Minutes of the State Board of Land Commissioners, December 17, 1917, 1–3.

32. "History: King Hill Project, Idaho"; Minutes of the State Board of Land Commissioners, December 17, 1917, 2–3; *Gazette*, July 27, August 24, December 7, 1917, February 15, March 22, 1918; Groefsema, 373–76.

33. *Salt Lake City Tribune*, June 27, 1922 (qtns.). "Annual Project History [for 1921],"155, roll 59, and "Data for Committee of Special Advisors" (1923), B-7, roll 62, in "Project Histories."

34. "History: King Hill Project" (1919), 32–33, 178–79 (qtns.), roll 58, and "Annual Project History [for 1921]," 15, roll 59, in "Project Histories."

35. "Data for Committee of Special Advisors," B-7, B-8; F. E. Wilson to Borah, January 15, 1929, Borah Papers; Groefsema, 371.

36. "Digest of Transcript of Hearings at Salt Lake City Convention—Special Advisors on Reclamation" (n.d.), 9, Garfield Papers; Resolution of Board of Directors of King Hill Irrigation District, November 17, 1923, in "Project Histories," roll 62.

37. Elwood Mead to Louis C. Cramton, August 14, 1925 (copy), and "Digest of Transcript of Hearings," 9; "Annual Project History [for 1923]," in "Project Histories," roll 60; *Twenty-Second Annual Report of the Bureau of Reclamation, 1922–1923* (Washington, DC: U.S. Government Printing Office, 1923), 13, 62.

38. Ray P. Teele, *The Economics of Land Reclamation in the United States* (Chicago: A. W. Shaw, 1927), 212.

39. *Idaho Statesman*, July 26, 1921; message from president on federal reclamation in 68th Cong., 1st Sess., 1924, S. D. 92, 210 (Serial 6238).

40. Letter from the secretary of the interior regarding federal irrigation projects in 69th Cong., 1st Sess., 1926, H.D. 201, 22–24 (Serial 8576); John W. Haw and F. E. Schmitt, *Report on Federal Reclamation to the Secretary of the Interior* (Washington, DC: U.S. Government Printing Office, 1935), 40; William E. Warne, *The Bureau of Reclamation* (New York: Praeger, 1973), 64, 68. King Hill settlers finally received federal patents to 13,702 acres by meeting Carey Act requirements for Irrigation of the land (Williams, 50).

41. *Idaho Statesman*, February 9, 1918.

CHAPTER 7

A "New West" Reclamation Tragedy: The Twin Falls-Oakley Project in Idaho, 1908–1931

At the turn of the century, utopian-minded writers and publicists proclaimed in moving prose the attractions and opportunities abounding in the "New West," that vast arid domain lying west of the Rockies which earlier pioneers had bypassed. Dedicated conservationists joined the chorus, declaring that a new era indeed had dawned, and cited the dozens of federal and state reclamation projects being launched under the Newlands and Carey Acts to reclaim these ample lands in that region. One of the areas pictured as potentially fruitful was the immense Snake River Plain in southern Idaho. Here, in 1908, the Twin Falls-Oakley Company started a reclamation project that within a decade clearly mirrored the difficulties that both entrepreneurs and settlers faced in their search for an Eldorado in a fragile land whose image had been grossly distorted by irresponsible promoters.

Nineteenth-century families traveling west on the Oregon Trail dreaded the part of their trip that began at Fort Hall in Idaho. Here they left behind cool, grass-covered shelters nestling at the foot of towering mountains, and entered the Snake River Plain, which stretched about 300 miles from Fort Hall to Farewell Bend on the Oregon line. Mostly a hot, uninviting desert, this region was covered with straggling sagebrush and strewn with flinty outcroppings of volcanic rubble that impeded travel. Fierce winds frequently lashed the plains, raising great clouds of soil particles that pelted hapless travelers. To avoid these conditions, westbound pioneers often detoured to the southward shortly after leaving Fort Hall, and sought trails into the foothills and mountains, which were crisscrossed by small rivers and creeks that rose in the high Idaho and Nevada mountain ranges to the south. The

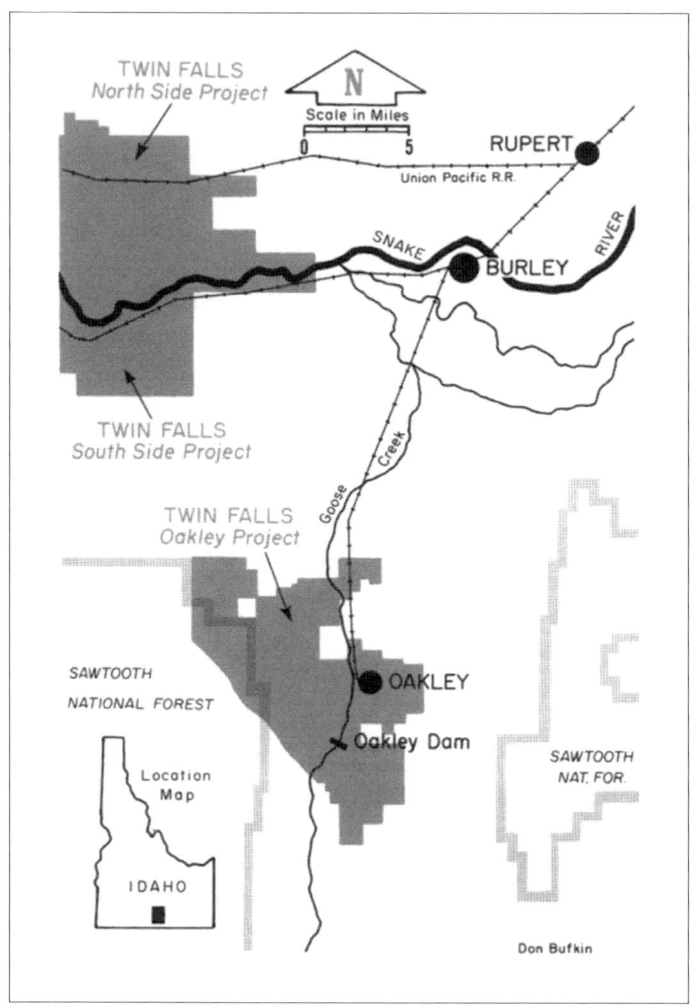

Carey Act reclamation projects. *This map by Don Bufkin originally appeared with this article in* Arizona and the West *20 (Spring 1978).*

highlands provided the caravans with water, forage for livestock, and welcoming shade from cottonwood groves clustered along the streams. By choosing this alternate route, pioneers avoided nearly one-half of the arid Snake River Plain and could rejoin the Oregon Trail in south central Idaho.

Many emigrants found desirable homesites along this alternate route, and a few abandoned their plans for continuing on to the Pacific Coast. In the 1870s, a handful of families deserted west-bound wagon trains and established farms along Goose, Birch, and Basin Creeks. Settlers from Utah joined them, creating more farms near these watercourses and founding villages at Oakley, Alma, Marion, Churchill, and Butte. Oakley grew the most rapidly and emerged as the hub of economic and cultural activities for the region.[1]

By 1900 settlers were drawing water from Goose, Birch, and Basin Creeks for irrigation and had placed 8,000 acres under cultivation. They planted principally hay and grains and, as soon as possible, accumulated flocks of sheep and herds of cattle. The hay and grain sustained the livestock during the winters, while untilled lands in the valley bottoms and the surrounding hills supplied space and forage during the grazing season. Depending on the livestock industry for an economic base, the settlers built a stable economy, developed a satisfactory mode of life, and created pleasing social institutions and surroundings.[2]

Although the village of Oakley prospered, the surrounding region in south central Cassia County grew slowly. The area was isolated geographically and little affected by the economic developments that drastically transformed large sectors of southern Idaho during these years. The Oregon Short Line Railroad, completed during the 1880s, provided a major East-West transportation link across the Snake River Plain, but it was located too distant to exercise much impact on Oakley and its environs. Furthermore, the building of the Utah & Northern Railway promptly captured that segment of the Utah-southern Idaho commerce formerly carried by stagecoach and wagon through Oakley.

More importantly, the Oakley region in the 1890s seemed to be overlooked when both federal and state agencies, enamored with the New West craze, authorized the construction of immense reclamation projects on the empty deserts in southern Idaho. Here the federal government designed and administered the Payette-Boise and Minidoka projects, and under the provisions of the Carey Act of 1894 it conveyed to the State of Idaho 1,000,000 acres where forty-two additional projects ultimately were started. As required by the Carey Act, the state contracted with private entrepreneurs to build irrigation works necessary to reclaim the land. Once the project was started, the state advertised land at the statutory price of fifty cents per acre. In southern Idaho, the new projects attracted thousands of newcomers who scurried to the newly-opened section with plans to build personal "Edens" overnight. Economic and social "progress" tended to be measured by the number of new farms and towns rising on the desert, and by the soaring values which Eldorado-seekers and land speculators placed on the reclamation tracts.

In 1908 the tranquility of the Oakley region was suddenly broken by the announcement that William and James Kuhn, Pittsburgh entrepreneurs, were planning to build a reclamation project in that locale. Owners of the American Water Works and Guarantee Company of Pittsburgh, Pennsylvania, and holders of extensive investments in banking and coal enterprises, the Kuhns were looking for additional places to earn profits. By 1908, they had founded electrical generating enterprises in Idaho and became developers of Carey Act reclamation projects there, most notably the Twin Falls North Side Project. Those operations begun, the Kuhns deemed the area around Oakley as a potential site for another development. Accordingly, the Kuhns invited the region to share the "progress" neighboring projects enjoyed, and formed the Twin Falls-Oakley Land and Water Company to achieve their goals.[3]

The Twin Falls-Oakley Company soon unveiled its plans. First, the Kuhns would build an earthen dam across Goose Creek to capture all runoff water from the stream and its tributaries. From this reservoir would run an intricate web of canals and laterals. According to their engineers, the reservoir would impound enough water to irrigate about 45,000 acres of land, each year supplying 1.5 acre-feet of water per acre. The Kuhns submitted these proposals to the Idaho Board of Land Commissioners (commonly known as the Land Board), the agency that administered Carey Act projects in the state, and requested permission to begin the irrigation system.[4]

The Idaho Land Board speedily approved the Kuhns' plans. This may have been due to the desire by the Utah-Idaho Sugar Company to obtain portions of the reclaimed land for extensive sugar beet farms. Accordingly, the board "segregated" 43,893 acres for the project, some 8,000 acres of which were privately owned by farmers. Calling the reclamation enterprise the Twin Falls-Oakley Project, the board signed a contract with the Twin Falls-Oakley Company on August 12, 1909. By the agreement, the corporation would construct irrigation works capable of servicing the segregated region. In return, it could sell water rights and services to farmers, pocketing the difference between its construction costs and charges to those occupying the lands.[5]

Goose, Birch, and Basin Creek farmers strongly disapproved of the Twin Falls-Oakley reclamation project, in which they were involuntarily included. Particularly vexing was the anticipated loss of untilled public lands surrounding their farmsteads where they had long grazed livestock and which they claimed by right of occupancy. Securing other grazing areas in the vicinity would be difficult, and Federal Forest Reserves and state-owned grazing lands were distant from their farms. Moreover, Forest Reserve grazing permits and leases on state-owned lands were monopolized by large sheep and cattle companies. The more pessimistic Goose Creek farmers speculated that the Oakley Project would ultimately force them to dispose of their herds of livestock. In that event, they would become dependent upon income from row crops such as potatoes and sugar beets and, perhaps, dairying. The prospect of performing stoop labor to operate potato and sugar beet farms was exceedingly distasteful. Dairying seemed almost as onerous.

Agents of the Twin Falls-Oakley Company had appeared at the gates of the farmers in the Goose Creek area as early as 1908. Preaching a gospel of quick riches by reclamation, they cited soaring land values on other projects, notably the prosperous Twin Falls South Side development, and urged the farmers to forfeit their "prior" rights to Goose Creek water. In return for these water rights, which the farmers (or their predecessors) had obtained from the state, the company agents offered a contract to supply 1.5 acre-feet of water yearly for each of their acres. This water, they promised, would be delivered promptly by an elaborate canal system. To make the trade more attractive, the agents also would give the farmers shares of stock in the Twin Falls-Oakley Company.

Company agents cajoled the Goose Creek farmers, too, with arguments that their own welfare was best served by cooperating. Among other reasons, the agents contended, the Twin Falls-Oakley Company was required by the Idaho Land Board to build and operate a railroad linking the Oakley Project with the burgeoning Snake River valley towns which the Oregon Short Line Railroad served. Surely the established settlers needed a railroad tie to the rest of Idaho, and access to out-of-state markets which the Oregon Short Line provided. The agents grossly exaggerated the benefits to farmers, it later developed, for the company operated the railroad only until 1915. Despite farmer protests, the Land Board never compelled the company to restore railroad service to the project.[6]

Goose Creek farmers rejected the agents' claims and scoffed at the get-rich-quick theories. They quickly pointed to shortcomings in the company's reclamation plans. They particularly questioned the proposal to supply water to five times as much land as was currently irrigated. Company calculations on the amount of Goose Creek water that could be impounded during the spring runoff and made available for summer irrigation also seemed misleading.

The farmers were powerless, however, to block the Oakley Project. They had no voice in determining the ultimate usage of the public lands adjoining their private holdings within the segregated area, nor could they prevent construction of the reservoir, for Goose Creek waters were not totally theirs. Moreover, the farmers found few allies to help them resist the project. A majority of the business, civic, and religious leaders at Oakley sided with the company. Supporters of the company argued that the farmers erred in claiming that Goose Creek waters were insufficient for so large a reclamation project. William T. Jack, president of the Cassia Stake of the Church of Jesus Christ of Latter-day Saints, marshalled the most persuasive rebuttals to the farmers. He proved that several irrigation projects in Utah succeeded when farmers shared sparse water supplies under a water rotation system. Also he accused Goose Creek farmers of irrigating their lands too heavily, thus projecting water needs far in excess of the real requirements for successful farming on the Oakley Project.[7]

Unable to keep an unwanted reclamation project from their backyards, most of the farmers capitulated. In August of 1908, they exchanged their prior water rights for Twin Falls-Oakley Company stock certificates and water contracts. The only holdouts were a handful of Goose, Birch, and Basin Creek stock raisers and the Vineyard Land and Stock Company, a major cattle ranching corporation with operations in southern Idaho and northern Nevada. Both eventually resorted to litigation and compelled the Twin Falls-Oakley Company to share Goose Creek water with them.[8]

The Oakley Project was initiated with high expectations. Financiers, small investors, and speculators in reclamation enterprises expressed lively interest. The Kuhns, their business associates, and other purchasers of Twin Falls-Oakley Company bonds risked large sums, variously calculated at between $1,500,000 and $2,250,000, in the venture. All anticipated handsome returns when project settlers fulfilled their water contracts with the company. Settlers were lured easily, for project backers and boosters scattered large amounts of tantalizing publicity catering to dreams of building Edens in the New West.[9]

The Oakley district was opened on September 23, 1909, and within a short time settlers had applied for 15,000 acres of state land, paying to the Idaho treasury the required fifty cents per acre. A majority also signed water contracts with the Twin Falls-Oakley Company. Many of the newcomers had left behind comfortable Middle West surroundings and now faced an uncertain future. Some were lucky. For instance, one individual purchased land and signed a water contract, and obtained work as proprietor of a general store at Marion to make ends meet until he could get his land cleared and a crop started. He recounted: "I am an Ohio clothing man. Came here for health and investment and will stay by it and encourage others to do so if we get what we are entitled to." Within a few more years, land seekers purchased another 15,000 acres and signed water contracts, leaving only about 5,800 acres still unoccupied. Unfortunately, some entrymen never fulfilled the Carey Act requirement that to obtain a patent they had to place at least one-eighth of their lands in cultivation. By 1916, approximately 11,000 acres within the project were unpatentable.[10]

Despite the enthusiasm of investors and settlers, misfortunes beset the Oakley Project from the first. As neither careful soil studies nor a systematic gathering of geological data preceded the construction, the company experienced difficulties in placing the canal and laterals properly and in building them to hold water in sandy areas. Most importantly, the construction of the Goose Creek dam took far longer than was anticipated, because of engineering miscalculations and mishaps at the site. The Twin Falls-Oakley Company finally began water delivery in 1913, but a majority of the settlers on the project had to wait another two years for water, as the irrigation system was not fully operational until the Goose Creek dam was completed and spring and summer runoff filled the reservoir in 1915. These delays caused years of deprivation for newcomers to the Oakley Project.[11]

In contrast to the plight of the newcomers, many of the "Old Settlers" were unaffected by the delays. A group of about 150 residing along Goose Creek continued to farm in their accustomed manner. Water was at their doorsteps, and they could readily raise the grain and hay crops which sustained their 5,000 head of livestock during the winters.[12]

Forced to wait for water delivery, newcomers to the Oakley district slackened the pace of clearing their acres and sought employment in the mines at nearby Vipont, in railroad construction, and in other wage-earning pursuits. In so doing, they created problems for themselves, for Carey Act land cultivation requirements were explicit and inescapable; the penalty was unpatentable land. Others became tired of waiting and drifted away in disgust, refusing to endure more years of disappointments. The remaining settlers impatiently struck back at the Twin Falls-Oakley Company by refusing to remit water payments, canal maintenance assessments, and certain petty fees stipulated in the fine print of their generally unread water contracts. J. H. Puelicher (president of the Twin

Falls-Oakley Company since 1913) warned settlers, in vain, that they were "in default and liable to all default penalties."[13]

More distress loomed for the project. In 1911 Eastern investors began complaining about securities issued by certain Western irrigation enterprises. They particularly questioned the operations of the American Water Works and Guarantee Company and other Kuhn-controlled corporations, and in 1913 forced all the Kuhn corporations into bankruptcy and the hands of a receiver. A group of bondholders organized a committee to protect the investors' interests. The committee quickly revamped the Twin Falls-Oakley Company, filed as security the water contracts held by the firm, and informed Oakley Project settlers that contracts would be enforced. The bondholders regarded the water contracts as basic to their financial salvation.[14]

Alarmed by bondholder threats, the settlers organized to defend themselves. They created the Oakley Canal Users' Association, and sought support from the Idaho Land Board. Many of the farmers also backed the gubernatorial candidacy of Moses Alexander, hoping that Democratic victories in the 1914 state elections would produce a Land Board sympathetic to their side. However, the bondholders proved less demanding than the settlers had anticipated. Digging deeply into their own pocketbooks, the bondholders raised more funds in order to ensure completion of the Goose Creek dam and other needed irrigation works. They also decided that only prosperous farmers could be held to the water contracts.[15]

The bondholders' committee and the Canal Users' Association bargained amicably and hammered out the so-called "Extension Agreement of 1914." Most of the Oakley Project residents approved the accord. By the agreement the Oakley Company would issue new water contracts which obligated farmers to place at least 50 percent of their lands under cultivation within three years. In return, the company postponed collection of water payments until 1919. In the interim, it required only remittances of interest (at 6 percent per annum) on the principal. The company also made concessions on several sore points, notably annual canal maintenance fees. With this compact, both settlers and bondholders looked ahead to happier times.[16]

But conditions did not improve. Little snow fell during the winters of 1915 and 1916, the Goose Creek Reservoir never filled, and the company could only deliver a maximum of one acre-foot of water to each farmstead acre. Citing their water contracts, indignant settlers demanded the 1.5 acre-feet of water to which they were entitled, but to no avail. They not only were forced to receive less but were compelled to endure an irksome rotation system during the hot summers. The rotation system, devised and enforced by the Twin Falls-Oakley Company, supposedly rationed water equitably, but farmers insisted that 1.5 acre-feet of water was a minimal requirement for successful cultivation. The friction increased. Rumors spread that the company played favorites, giving a few settlers plenty of water at the expense of the multitude.

At the same time, Old Settlers rued their rashness in surrendering water priorities in 1908 to the Twin Falls-Oakley Company. They demanded full allocations of water and became restive on hearing disturbing gossip that the Kuhn interests earlier had paid $50,000 to prominent civic and religious leaders for whipping the Old Settlers into line. Denied more water and their patience exhausted, the Old Settlers sued to regain their water priorities. But the Idaho Supreme Court disappointed them. In 1916 the court held that Old Settlers rights were "identical with the rights of the [other] entrymen." Old Settlers appealed the issue to the federal courts, but again were denied redress.[17]

Meanwhile, a group of farmers had appealed to Governor Alexander. They asked that he compel the company to honor existing water contracts and make amends for crop losses, halt further settlement on the project, and find permanent solutions to the local water problems. Alexander responded by holding hearings. Then he persuaded the Twin Falls-Oakley Company and Canal Users' Association to negotiate. But the talks were unproductive. Association spokesmen charged that the company was totally unresponsive, refused to discuss reimbursement for 1915 crop losses, and selfishly upheld the financial interests of the bondholders. Moses Smith (an association spokesman) reported: "The poor bondholder is about all the Company people can see…[and] if the bondholders are worse off than the settlers, I think the various States will take care of them in there [sic] poor houses." When the talks ceased, most of the farmers again refused to pay monies due the company.[18]

When the 1916 irrigation season began, the Twin Falls-Oakley Company retaliated, withholding water and placing liens on farmlands. The farmers again went to the courts for defense. They sued the company and obtained a court order compelling it to supply water until the Idaho Supreme Court adjudicated the dispute. The company, however, extracted an extra pound of flesh by supplying water mostly during the autumn, instead of July when thirsty crops most needed it. Farmers calculated crop damages at $100,000 that year.[19]

Although punished economically, the water users prevailed in the courts. In July of 1916, the state supreme court ordered the company to resume water delivery to the project farms, holding that the company's acts conflicted with good public "policy," Idaho statutes, and the Idaho Constitution, which required that "public waters" be used for the "highest beneficial use." The court also cancelled most of the company liens on farmlands. According to the justices, Carey Act lands on the Oakley Project were not as yet patented and were, therefore, state or federal property on which the Twin Falls-Oakley Company could not place liens. The company could secure redress by filing liens on 1916 crops. Company officials were not enthusiastic about this suggestion, because crops that year were not plentiful.[20]

Feuding between the farmers and the Twin Falls-Oakley Company had resolved none of the project's fundamental problems. The disastrous 1915 water shortages, and the fight over deliveries in 1916, clearly demonstrated that the Oakley Project was too large, and had to be reduced to an acreage consonant with available Goose Creek water. Harsh as that conclusion seemed, no other option was in sight, for the project was too distant from the Snake River or another major water source. Their earlier judgment of Goose Creek's irrigation capacity vindicated, the Old Settlers now called for drastic slashes in the total authorized acreage. They persuaded most of the project entrymen to support a reduction from 35,800 acres to 15,000 to 18,000 acres.[21]

The Twin Falls-Oakley Company faced major financial losses if project acreage were reduced. It quickly countered with a proposal to reduce the project to 28,000 acres, the reduction to be effected by eliminating approximately 10,000 acres already abandoned by discouraged entrymen. In this way few water contracts would be cancelled, and bondholders would realize minimal losses. Idaho state officials generally backed the company plan. State Engineer J. H. Smith (the Land Board's principal technical advisor) recommended reductions to 26,800 acres, while the Irrigation and Drainage Code Commission deemed 26,730 acres suitable.[22]

The farmers hastily rejected the company proposals and appealed anew to Governor Alexander. Although the governor was a Democrat and the Land Board predominantly Republican, they asked that he and the agency resolve this new difference of opinion. Alexander opened hearings at Oakley in February of 1916, and the Idaho Land Board eventually announced a possible solution. The board decreed a project of 21,000 acres, and required that the Twin Falls-Oakley Company reimburse those who would be excluded from the project. Excluded entrymen would be paid for their improvements on abandoned lands, and would receive vacant lands within the reduced project acre-for-acre. These terms were conditional upon acceptance by the company and the settlers, but the board assumed that both groups, after prolonged grumbling, would ratify the terms.

The board immediately implemented its decisions. It ordered the state engineer to determine specific lands to be excluded, and announced that Idaho would be relieved of administrative obligations for the Oakley Project when the board's decisions became effective. In November of 1916, the board declared that the State of Idaho had divested itself of further involvement in the Oakley Project. It formally accepted the project as completed and ruled that the Twin Falls-Oakley Company had fulfilled all of its Carey Act contractual obligations. The board then took steps to secure land patents for settlers residing within the 21,000-acre project.[23]

The Land Board's 1916 solutions to the Oakley Project problems did not endure. Democrats elected to the agency in 1916, heeding protests from farmers and the Twin Falls-Oakley Company, abruptly cancelled all board actions that applied to the

Oakley Project and most of the other Carey Act reclamation projects in Idaho. They then asked Clay Tallman, United States General Land Commissioner, to conduct an investigation of all pending Carey Act projects in the state.

Tallman made a painstaking investigation, and his inquiries generated great dissatisfaction. Farmers complained that his fact-gathering and lengthy deliberations would cause more water-short irrigation seasons and more crop losses. Others regarded Tallman's time-consuming procedures as subjecting Oakley Project farmers to needless "uncertainty." A newspaperman commented on these objections: "Conditions are so unsettled on the Oakley Project that there is little activity manifest among the settlers. Before there will be any permanent improvements made or any really earnest efforts to build up the farms any further, the entrymen must be assured of water for their crops." Officers and bondholders of the Twin Falls-Oakley Company were equally disturbed. They feared that farmers, upset by uncertainties concerning the project, might produce few crops in 1917 and be unable to remit monies due the company. The officials also fretted about the Idaho Land Board being controlled by Wilsonian Democrats, who presumably sympathized with the underdog and would treat company interests lightly. Commissioner Tallman finally acted. He limited the Oakley Project to 20,500 acres. Subtracting 9,600 acres held by Old Settlers, he authorized the issuance of patents to the remaining 10,900 acres within the project.[24]

With some dismay, Governor Alexander and the Land Board quickly found that Tallman's decisions were not popular. The Twin Falls-Oakley Company flatly rejected so small a reclamation tract, while officers of the Canal Users' Association demanded further reductions in the size of the project. Pressured by both groups, Alexander firmly declined to travel either "road." For their part, those farmers who found their acreage now excluded believed their oxes unduly gored. Especially outraged was a group of Old Settlers who learned that their 1,800 acres were excluded; they immediately sought to save themselves by re-establishing their "prior" rights to Goose Creek water. Kenyon sector farmers also fought for re-inclusion in the project because their lands were, State Engineer

E. H. Hasbrouck admitted, "some of the best in the district." But Kenyon farmer pleas went unheard. Their lands were the most distant from Goose Creek and, therefore, were deemed the most difficult to supply with water. Tallman's decisions on Oakley Project reductions were appealed many times during the next few years by "excluded" farmers and irate Old Settlers. Its purse greatly lightened, the Twin Falls-Oakley Company repeatedly objected, too.[25]

Settlers on excluded lands who threatened lawsuits to recover their lands and water rights found that avenue closed in 1921. In that year a United States Court of Appeals upheld the legality of reclamation project reductions, under nearly identical circumstances, on the nearby Twin Falls-Salmon River reclamation project. In that case, *Twin Falls-Salmon River Land and Water Company v. A. E. Caldwell, et al.*, the court held that Carey Act land segregations could be increased only when additional water was available to reclaim lands. It confirmed that the General Land Office had the authority to determine whether lands were reclaimed within the meaning of the Carey Act and subsequent amendments. Excluded farmers fought back on other grounds, arguing that the construction of additional reservoirs on the Goose Creek watershed would provide storage for more runoff water. Hence, their lands should be restored to the Oakley Project. However, studies by the United States Bureau of Reclamation and the Idaho Department of Reclamation refuted the settlers' contentions.[26]

Although many farmers abandoned their excluded farmsteads, others stayed and farmed for a few years. The Oakley Company provided them water through the 1920 irrigation season. Persistent settlers actually occupied their farms until January of 1922, with state officials protecting their tenure by postponing formal relinquishment of excluded lands to the United States Department of the Interior.[27]

Excluded settlers also demanded compensation from Idaho. They claimed that the state owed them monies for past improvements on their lands, pointing out that Commissioner Tallman's mandates ignored the question of compensating those with excluded farms. Governor David Davis (in office since 1919) rejected most of these demands, but offered his help in mediating between aggrieved settlers and the Oakley Company. The Davis administration pressured the company to reimburse excluded settlers by returning a reasonable portion of the money which they had paid on their water contracts. Also Davis persuaded Idaho legislators in 1921 to pass a law refunding to the settlers the fifty cents per acre they had paid to the state treasury when they first occupied the land. There was grumbling about so little compensation, but no more state funds were forthcoming. Reflecting the indignation, one newspaper editor wrote: "While fifty cents an acre is better than nothing, it no wise repays those unfortunate settlers for their losses and suffering on those [excluded] lands due to the failure of the state of Idaho to determine the water supply before the [Carey Act] projects were approved."[28]

In 1921, excluded settlers believed they had one last chance for salvation. Reports circulated that there were immense petroleum deposits near Oakley, the boosters claiming that the area would supplant Tulsa as an oil center. But drillers found only traces of petroleum, while artesian water gushed from their wells. The artesian water sparked hope that mighty underground streams lay beneath the surface and could be tapped for irrigation. Extensive drilling disproved that theory. Their last hopes dashed, a majority of the excluded settlers abandoned their farmsteads. Only a few stayed, encouraged by a federal law which allowed them to keep their lands by filing for patents under the Homestead Act of 1862. Later, a persistent handful were more fortunate: in 1928 they won re-inclusion of their lands in the Oakley Project.[29]

In the meantime, confrontations between settlers and the Twin Falls-Oakley Company had continued. Water shortages developed in 1918 and became acute in 1919, with water deliveries dropping below contracted amounts. Angry settlers again struck at the company, refusing to make water payments, which had been suspended per the Extension Agreement of 1914, but were due in 1919. Quarrels over canal maintenance schedules, company policies on releasing water for livestock during the winters, and other management practices further exacerbated the differences.[30]

Some progress was achieved in resolving the latest disputes. In 1920, the Twin Falls-Oakley Company eliminated most of the friction over canal management practices. It transferred operation of the irrigation system to the Oakley Canal Company, a corporation in which settlers' representatives were accorded a voice. Formed in 1916, when the Land Board declared the Oakley Project as completed, the Canal Company had been forced to remain inoperative because of actions by state officials. For example, early in 1917, the Land Board had rescinded the 1916 certification of project completion, and the Carey Act prohibited settlers from administering an irrigation system prior to state approval. Idaho officials made the required certification for the Oakley Project in 1920.[31]

On the other hand, money disputes could not be resolved. Discussions between the Twin Falls-Oakley Company and the Canal Users' Association proved fruitless, largely because the association insisted that a final settlement on money issues be conditional upon the company equitably compensating settlers whose lands had been excluded from the project. Governor Davis and the Idaho Department of Reclamation strongly backed the association's position. The company balked at these demands, for it already had incurred substantial losses from the project reductions, which had caused the company to void its water contracts with the excluded settlers. Faced with the company's adamant guarding of its treasury, the settlers terminated the negotiations, vowing to seek "fair treatment" from the courts.[32]

Troubles between settlers and the Twin Falls-Oakley Company worsened significantly in 1921–22. Water supplies were plentiful and crops were abundant, but a nationwide depression in agriculture cut into the incomes of Oakley Project farmers. Consequently, many ignored their water contract obligations. Embittered that heavy crop losses in past years were still uncompensated, others considered this action fully justified. In response, the company filed liens and instituted foreclosure suits on the settlers' lands and water contract rights, believing that its legal contentions were strengthened now that settlers had ample water. The Canal Users' Association mobilized to block the liens and foreclosures. The officials demanded support from local non-farm residents and collected a sizeable legal defense fund. However, their efforts were unavailing. By August of 1923, the Oakley Company had carried foreclosure suits on about 3,000 acres and attached water rights "through the courts to decree and sale and in one case to deed."[33]

Particularly disheartening to settlers was the outcome of one lawsuit, *Twin Falls-Oakley Land and Water Company v. John H. Martens, William G. Moyes and John S. G. Walker*. Here, the plaintiff, seeking to foreclose on the land and water rights of the defendants, was rebuffed in the United States District Court. The court decreed that the foreclosures could be quashed if the defendants paid seven-ninths of the amount the company claimed in arrears. According to Judge F. S. Dietrich, the seven-ninths represented the fractional proportion of contracted water which the company delivered to the defendants during the years covered by this lawsuit. However, a court of appeals reversed portions of Judge Dietrich's ruling, and the United States Supreme Court refused to hear the case. The defendants' properties were foreclosed and sold at a sheriff's sale.[34]

Company foreclosures, coupled with farm losses from the prevailing depression, convinced project farm leaders that an immediate compromise with the company was necessary. At the outset, one faction loudly opposed an accommodation, arguing that the courts would ultimately help the settlers by dictating terms never attainable in negotiations with the company. But this view steadily lost support, particularly when the residents of the Churchill sector of the project publicly called for a settlement with the company. Moreover, the Idaho Department of Reclamation quietly but firmly interposed its opposition to further litigation in the matter. It suggested the formation of an irrigation district in order to eliminate the Twin Falls-Oakley Company from continued involvement in the project. The department argued that the sale of irrigation district bonds could fund a settlement with the company for its rights in the Oakley Project.[35]

In 1923 the farmers heeded the state reclamation agency's suggestions. They elected a special committee, headed by William T. Jack, and instructed the group to resolve the water crisis. The committee made the first overtures to the Twin Falls-Oakley Company, and in July they reached an agreement. The company would bow out of the Oakley Project as soon as an irrigation district was formed and reimbursement made for a portion of its investment. The reimbursement would be computed on the basis of the company claim to water contract payments on 21,057 acres. The amount was set at $70 per acre in certain instances, and on the remaining acreage at book value plus interest to April 1, 1923, less the amounts that settlers had remitted per their water contracts. The indebtedness to the company would be secured by notes and mortgages underwritten by Oakley Project landholders and the Oakley Irrigation District. Most of the debt would be liquidated from the proceeds of district bond sales over a period of twenty years.[36]

For numerous reasons, the 1923 pact proved unacceptable. It ignored the claims for compensation by settlers on excluded lands. It recommended intolerably high payments to the Twin Falls-Oakley Company and, most importantly, it angered the Old Settlers and a few others whose water contracts had been paid. Holders of paid-up contracts and Old Settlers alike supported an end to incessant battles with the company, but steadfastly resisted any settlement by which they shared the water contract obligations of others. The Settlers' Special Committee, which had been the architect of the 1923 pact, had second thoughts and joined the opponents of the agreement. In November of 1923, and again in April of 1924, settlers voted down the proposed irrigation district, thus rendering the 1923 agreement inoperable.[37]

Although the 1923 agreement was rejected, it contained suggestions for a way out of the morass enveloping the Oakley Project. Clearly the company's bondholders must temper their financial demands and absorb greater losses. Settlers without "paid-up" water contracts must shoulder the burden for paying whatever sums that might be needed to secure the exit of the company. Settlers excluded from the project had to accept the little help that state and federal legislation afforded them. Accordingly, negotiations were resumed in 1924, only to be shelved temporarily while a plan for bringing Snake River water from the distant Minidoka Dam was considered. Eventually the plan was rejected as too costly. Talks between the company and the Settlers' Special Committee then resumed, producing in 1928 an agreement which the company's bondholders and the various settler factions grudgingly accepted. Implementation of it was delayed, however, until the courts in 1931 concluded litigation concerning the legality of 1918 reductions of Oakley Project acreages.[38]

In the end, state and federal decisions halving the Oakley Project to about 21,000 acres prevailed, leaving many losers in the long struggle. Company bondholders eventually recovered less than 50 percent of their original investments. The Old Settlers irretrievably lost their prior Goose Creek water rights, and shared thereafter the adversity of their neighbors during dry seasons. Most of the settlers on excluded acres were compensated with fifty cents per acre paid to them by Idaho, and enjoyed the dubious privilege of homesteading land where water was unavailable. All Oakley Project entrymen wrote off past crop damages that had been incurred due to Land Board errors and the Twin Falls-Oakley Company's greed in launching a reclamation project without sufficient water resources. Consequently, all groups were contrite when the din of battle ended. None stepped forward to claim the victor's laurels. There were none to claim.

The Oakley Project in Idaho was unique among Carey Act reclamation enterprises. It was superimposed on an area populated by farmers who were satisfied with their lifestyles and enjoyed a prosperous stock raising economy. Suddenly and against their wishes they were catapulted into the New West's reclamation frenzy with its attendant hardships and frustrations. To their dismay they were forced to pioneer anew, this time alongside the Carey Act land entrymen who arrived after 1908. Both groups quickly discovered that New West pioneering at Oakley required constant dealings with greedy capitalists and suave public officials who employed highly sophisticated methods and harbored complex motives.

Bondholders and officials of the Twin Falls-Oakley Company were dominated, many settlers charged, by avarices equaling those of the Old West's beef barons and sheep kings. Old Settlers particularly resented their fate, and sometimes became nostalgic for the Old West's simplicities. Battling with marauding Indians and crowds of rowdy, fence-snipping cowhands seemed more even contests, than were conflicts with the Idaho Land Board and the Twin Falls-Oakley Company. Old Settlers and Carey Act entrymen alike doubtless wondered if the dreams and hopes of the New West were not built on quicksand.

This article originally appeared in *Arizona and the West* 20 (Spring 1978). Reprinted with permission.

Notes

1. M. D. Beal, *A History of Southeastern Idaho* (Caldwell, ID: Caxton, 1942), 189–90, 411n; Leslie L. Sudweeks, "Early Agricultural Settlements in South Idaho," *Pacific Northwest Quarterly*, XXVIII (April 1937), 149–50; *Oakley Herald* (Idaho), May 5, 1922.

2. J. W. Barber, "Cassia County Agricultural Agent Annual Report for 1924," November 30, 1924, in Annual Narrative and Statistical Reports from State Offices and County Agents, Idaho, 1913–1944, Microcopy T857, Roll 9, Records of the Federal Extension Service, Record Group 33, National Archives.

3. The Twin Falls-North Side Project was the largest of the Kuhns' reclamation enterprises. This project also experienced many serious problems until the 1920s.

4. James J. Stephenson Jr., *Irrigation in Idaho* (Washington: U.S. Department of Agriculture Bulletin 216, 1909), 46.

5. Water rights were sold initially to settlers for $40 an acre; by 1913 the price had been increased to what one irate farmer called "the outragous [sic] price of $65." Moses Smith to Moses Alexander, February 12, 1915, Moses Alexander Papers, Idaho State Archives [ISA], Boise. Settlers also were charged for maintenance and operation of the irrigation works.

6. W. L. Dunn to State Board of Land Commissioners, March 9, 1917; "Twin Falls Oakley L. and W. Co.," undated typescript, Alexander Papers, ISA. *Wendell Irrigationist* (Idaho), November 18, December 2, 1915. The Oakley Company offered farmers variable shares of stock; the offers were usually based on the numbers of miners' inches of Goose Creek water on which farmers had a filing. Sometimes the number of shares of stock became a negotiable issue between company and farmers.

7. William T. Jack to R. W. Faris, February 13, 1909, copy in *John Adams, et al. v. Twin Falls-Oakley Land and Water Company*, U.S. District Court (Boise), file S687, Federal Archives and Records Center [FRC], Seattle.

8. Mikel H. Williams, *The History of the Development and Current Status of the Carey Act in Idaho* (Boise: Idaho Department of Reclamation, 1970), 75. Dunn to Land Board, March 19, 1917, Alexander Papers, ISA. *Oakley Herald*, August 13, December 24, 1915.

9. Williams, *Carey Act in Idaho*, 69. J. H. Puelicher to Clay Tallman, December 12, 1917, Alexander Papers, ISA. *Oakley Herald*, June 19, 1914.

10. H. E. Smith to Alexander, January 5, 1915; Dunn to Land Board, March 9, 1917, Alexander Papers, ISA.

11. *Oakley Herald*, February 28, 1913.

12. Dunn to Land Board, March 9, 1917, Alexander Papers, ISA.

13. *Oakley Herald*, June 19, 1914.

14. Williams, *Carey Act in Idaho*, 75. *Oakley Herald*, June 19, 1914.

15. *Oakley Herald*, July 4, 1914; March 12, 1915.

16. Ibid., June 19, September 18, December 11, 1914. Prior to 1914, settlers were charged maintenance fees of thirty-five cents per acre per year. The Extension Agreement required them to pay only actual maintenance costs.

17. George A. Day to Alexander, January 27, 1916; Dunn to Land Board, March 9, 1917; W. C. Schlegel to Alexander, August 2, 1915, Alexander Papers, ISA. *Rome Adams et al. v. Twin Falls-Oakley Land and Water Company*, 29 Idaho Reports 365. *Oakley Herald*, July 17, 1917; *Burley Bulletin* (Idaho), March 28, 1919, and February 17, 1920; Boise *Idaho Statesman*, December 3, 1920. Order of Judge Charles C. Cavanah, February 13, 1928, U.S. District Court (Boise), file S687, FRC, Seattle.

18. *Oakley Herald*, July 23, 1915. Alexander to Faris, July 18, 1915; Moses Smith to Alexander, July 24, 1915, and April 22, 1916, Alexander Papers, ISA.

19. Dunn to Land Board, March 9, 1917, Alexander Papers, ISA. The company invoked Section 6 of the 1914 contracts with settlers: "It is agreed that no water shall be delivered to the purchaser from said irrigation system while any installment of principal or interest is due or unpaid from the purchaser to the Company." Cited in *Oakley Herald*, May 26, 1916.

20. *Rome Adams et al. v. Twin Falls-Oakley Land and Water Company*, 29 Idaho Reports 369, 374–76. *Oakley Herald*, June 21, 1916.

21. "Twin Falls Oakley L. and W. Co."; Moses Smith to Idaho Drainage Code Commission, November 13, 1915, Alexander Papers, ISA.

22. "Report of the Idaho Irrigation and Drainage Code Commission," December 1, 1915; J. H. Smith to Land Board, March 26, 1916, Alexander Papers, ISA. *Oakley Herald*, December 10, 1915.

23. "Twin Falls Oakley L. and W. Co."; Puelicher to Tallman, December 12, 1917, Alexander Papers, ISA. *Oakley Herald*, February 16, March 3, 1916. Settlers were in no hurry to obtain land patents, because they feared that the company could place liens upon their land during future disputes.

24. "Twin Falls Oakley L. and W. Co."; Tallman to Land Board, August 13, 1918, Alexander Papers, ISA. *Oakley Herald*, March 16, October 19, 1917.

25. Puelicher to Tallman, December 12, 1917; Dunn to Alexander, June 5, 15, July 11, 18, August 19, 29, 1918; Alexander to Dunn, June 21, 1918; E. H. Hasbrouck to Alexander, August 27, 1918, Alexander Papers, ISA. Tallman ordered that lands excluded from the project be returned to the public domain.

26. *Oakley Herald*, November 28, 1919; September 24, 1920; March 4, 1921. Boise *Idaho Statesman*, March 14, 1921. W. G. Swendsen to D. W. Davis, May 9, 1919; Davis to L.B. Compton, July 16, November 26, 1921, David W. Davis Papers, ISA.

27. *Oakley Herald*, April 22, 1921.

28. Compton to Davis, July 11, 1921; Davis to Compton, July 16, 1921; Swendsen to Davis, November 16, 1921, Davis Papers, ISA. House Bill 246, *Session Laws of Idaho* (1921), 83–86; *Seventh Biennial Report of the Department of Reclamation, 1931–1932* (Boise: Idaho Department of Reclamation, 1932), 34. *Burley Bulletin*, March 10, 1921.

29. *Oakley Herald*, June 11, July 23, 30, August 13, 1920; February 11, March 11, December 23, 1921; December 29, 1922. *Fifth Biennial Report of the Department of Reclamation, 1927–1928* (Boise: Idaho Department of Reclamation, 1928), 15.

30. See also note 16. The Extension Agreement obligated settlers only to pay interest on the outstanding principal of their water contracts until 1919. The principal was payable in several annual installments beginning in 1919.

31. *Oakley Herald*, February 6, 1920.

32. Ibid., April 29, 1921. Swendsen to Davis, November 16, 1921, Davis Papers, ISA.

33. *Oakley Herald*, December 8, 1922, August 3, 1923.

34. *Wendell Irrigationist*, December 18, 1919. *Oakley Herald*, March 13, 1922.

35. *Oakley Herald*, April 14, 1922. Swendsen to C. C. Moore, July 20, 1923, Charles C. Moore Papers, ISA. *Third Biennial Report of the Department of Reclamation, 1923–1924* (Boise: Idaho Department of Reclamation, 1924), 7.

36. *Oakley Herald*, July 13, August 3, 1923.

37. Ibid., August 3, September 14, 21, October 26, November 9, 1923; *Burley Bulletin*, November 29, 1923, May 8, 1924. Dunn to Swendsen, January 22, March 29, 1924; B. F. Wilson to Swendsen, March 1, 1924, Oakley Irrigation District Records, File 107A, Idaho Department of Water Resources, Boise.

38. *Oakley Herald*, November 21, 1924. "The Oakley Valley Story," *Cassia County's 46th Annual Fair and Rodeo* (Burley: County Fair Board, 1956). *Fifth Biennial Report of the Department of Reclamation, 1927–1928*, 27; Williams, *Carey Act in Idaho*, 75. As of 1928, 18,894 acres were irrigated within the Oakley Project. *Report on Irrigated Area and Classification of Water Rights in Idaho* (Boise: Idaho Department of Reclamation, 1928), 6.

CHAPTER 8

Yellowstone National Park, Jackson Hole, and the Idaho Irrigation Frontier

Federal scientists in the 1890s classed as irrigable about 5,000,000 acres in Idaho's Snake River Basin. About two thirds of this irrigation empire was situated in the easternmost parts of Idaho, abutting Wyoming, and in the state's south-central areas bordering on Utah and Nevada. Bisected by the Snake River, this eastern half of the Snake River Basin stretched from the Snake River's headwaters in Jackson Hole, Wyoming, to the southside area surrounding Twin Falls. In this area, it became clear early in the twentieth century, eastside and southside dwellers could build their irrigation empire on roughly 3,000,000 acres; but reclamationists first needed, as economically as possible, to store much of the Snake River system's water in reservoirs for summer irrigation purposes. They might also be forced to augment Snake River water with water diverted into the basin from outside sources. An irrigation scheme popular with the eastside/southside crowd because of its relative simplicity and purported economy entailed freewheeling use of the water resources and reservoir sites at Jackson Hole and the diversion of water from Yellowstone National Park over objections from the federal government and conservationist lobbies. Disputes over so utilizing Jackson Hole and Yellowstone water recurred over the decades from 1900 to 1940.

This struggle was so lengthy because schemes for diverting Yellowstone water to the Snake River Basin became rooted in irrigationists' thinking virtually to the point of being their fondest panacea. Only with difficulty could state officials, federal authorities, and conservationist forces convince the irrigationists to accept alternatives. This struggle for water resources revolved primarily around Yellowstone Lake and Cascade Corner in the southwestern corner of Yellowstone National Park.

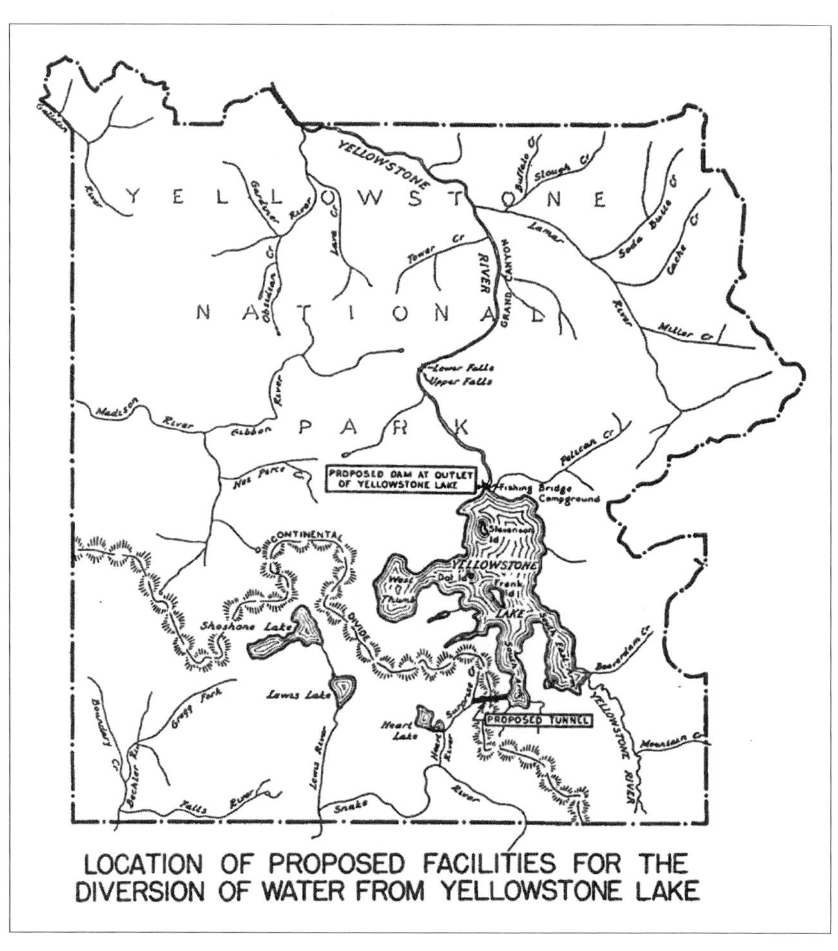

This map, from "Statement on Diversion Proposal Affecting Yellowstone Lake, Yellowstone National Park, As Embodied in S. 3925 and HR 10489, 75th Congress," appeared in *Planning and Civic Comment*, January–March 1938, and was reprinted with this article in *Idaho Yesterdays 43, no. 4 (Winter 2000).*

In such conflict over whether irrigationists should win access to Jackson Hole and Yellowstone water, the salient features of the fighting centered from one perspective on conservationists' idealism vis-à-vis the opportunism of irrigationists. The latter put inexpensive irrigation procedures ahead of saving the superlative scenic and wildlife values of Jackson Hole and Yellowstone National Park. Such an interpretation of what was at the core of these struggles also reflected how grander public-policy issues had become a part of clashes between two of the twentieth-century West's major initiatives, reclamation and scenic preservation. However, important as the preservation-versus-reclamation dichotomy was in understanding the fight over use of Jackson Hole and Yellowstone National Park for irrigation purposes, the struggles could be read more fundamentally as a case study in western irrigation politics.

As had happened so often in western irrigation politics, the controversy over Yellowstone and Jackson Hole water grew in complexity because it soon involved many private constituencies and public entities. At first, Idaho's eastsiders and southsiders alternately put forward rival schemes but sided with each other in fights with national-park defenders—fights that never made it to the courts. On both sides of the eastside/southside fence, tension also mounted between camps that relied on free enterprisers for reclamation of arid land and those who benefitted from the federal Reclamation Service's working. Eventually, the distinctions between eastsiders and southsiders seemed to fade. In fact, they were overshadowed rather than subsiding when interstate issues between Idaho and Wyoming and intradepartmental rivalries between the Reclamation Service and the National Park Service made it more difficult to resolve disputes about the use of Jackson Hole and Yellowstone water for irrigation.

These fights also fit into other molds common to western irrigation politics. For example, much as in other states, it became essential in the Snake River Basin to rectify by political means the misjudging by early irrigationists of the engineering and economic realities of grand-scale irrigation. Aside from relying on inadequate financing, ignoring settler welfare, and gainsaying other conditions, Idaho groups had underestimated the water resources that were needed to irrigate their land. In a single year, farmers at one tract consumed 42,000 acre-feet of water above what their water contracts granted them. Still, an irrigator complained, "the [irrigation] system does not water thirty-five percent of the land."[1]

As early as 1890, U.S. Geological Survey scientists maintained that water from Yellowstone National Park and Jackson Hole should sustain irrigated farming in Montana and Idaho. Montana irrigationists, they judged, could best utilize Yellowstone Lake, the grandest of several lakes in the national park of the same name; the same reclamationists expected water from Jackson Lake, the largest lake in Wyoming's Jackson Hole, to be consumed in the eastern half of the Snake River Basin. Geological Survey engineers calculated that because Jackson Lake was situated near the head of the Snake River water system, irrigation would come dirt cheap if 500,000 acre-feet of the lake's water were used in Idaho. The lake, which sometimes held as much as 9,000,000 acre-feet over a year's span, might serve additional irrigation tracts at higher costs. Costs would be elevated to pay for a dam at the lake's outlet.[2]

At the U.S. Reclamation Service, a new agency charged with hastening western land reclamation after 1902, director Frederick Newell and his engineers seized on the Geological Survey's findings on Jackson Lake. The Survey's hydrological surveys proved to them that the Reclamation Service's 130,000-acre Minidoka Project, a new venture on the south side of the Snake River Basin, could be irrigated satisfactorily by converting Jackson Lake into a reservoir and building another catchment near the project. Both reservoirs were essential. The catchments were needed to impound spring runoff water for the project because early pioneers and new irrigators at the Twin Falls tract had preempted the natural summertime flow of the Snake.[3]

By Newell's plans, his agency would altogether reclaim about 500,000 acres at Minidoka and the Boise Project—a larger federal project on Idaho's western edge—under the authority of the Reclamation Act of 1902.[4] Newell also envisioned his agency building its own Dubois irrigation project on an empty but readily irrigable expanse of over 200,000 acres at the east side of the Snake River Basin. Such a possibility figured in Reclamation Service planning until, in 1904, the bureau suspended Dubois for economic and political reasons. Two years later, it abandoned the project.[5]

Newell and his agents also reviewed the possibilities of supplying water at Minidoka from sources in Yellowstone National Park. Aside from large reserves at Yellowstone Lake, federal surveyors reported that nearby at Shoshone Lake were "splendid facilities" with which to collect another 79,884 acre-feet of water for irrigation. Also in the same sector of the park, they calculated, Lewis Lake held lesser amounts of water that could be utilized for Minidoka wet farming.[6] However, Reclamation Service administrators finally ruled out using these sources for two reasons. First, transferring Yellowstone water across mountains to the Snake River Basin seemed likely to cost the government more than relying on irrigation works at Jackson Hole. Second, federal hydrologists believed that by damming the outlet of Jackson Lake and emptying water from the lake into the Snake River, they could lower water costs to Minidoka irrigators to one dollar per acre—about half the price of water delivery on many non-government projects.[7] In Jackson Hole, Reclamation Service surveyors also noted, water for irrigation purposes was available at Two Ocean, Emma Matilda, Leigh, and Jenny Lakes.[8]

Implementing the decision to dam Jackson Lake, the Reclamation Service blocked the lake's outlet in 1907 and improved the works in 1911 and 1913. Minidoka farmers prospered because the new irrigation works ensured them sufficient water in the 1910s. The federal bureau also hastened arid-land reclamation at non-government projects in southern Idaho when it increased the storage at Jackson Lake to 847,000 acre-feet by raising the dam four years later. Federal authorities sold equities to thousands of acre-feet of federally impounded water at Jackson Lake to two big Twin Falls irrigation tracts. At those projects, which contained almost half of the 800,000 acres in Idaho that were reclaimed under provisions of the federal Carey Act of 1894, irrigators believed that their purchase of water had supplemented their rights to natural-flow water in the Snake River to the point that they were immunized from water shortages.[9] An outsider even classed Jackson Lake as "the Savior of the Snake River Valley."[10] In the network of independent canal companies and publicly constituted irrigation districts that controlled irrigation in Idaho's eastern counties bordering on Wyoming, several of the more affluent entities also purchased equities in 100,000 acre-feet of Jackson Lake water in 1917.[11]

Except in this last instance, irrigationists in the eastside counties had resisted any dealings with Reclamation Service authorities at Jackson Lake. Dominating a six-county region situated above the American Falls of the Snake River, an area sometimes called the Upper Snake River Valley, these farmers had relied historically on the Snake River's natural but typically low water flow in late summer. About 114,000 of their acres were watered inadequately, but toward the end of the irrigating season, conditions like quirky weather compounded their water shortages.[12] Nonetheless, they still shied away from buying water equities at Jackson Lake. Resisting partly on grounds that the Reclamation Service overcharged nongovernment water users for water equities at the lake, Upper Snake Valley irrigationists also decried the federal agency's practice of compelling non-government entities to pay in advance annually for government-impounded water at the lake. Moreover, many of the same eastsiders deemed it unrealistic to expect Jackson Lake water for themselves over the long term. Already Wyoming authorities had thundered at each expansion of Jackson Lake irrigation works because using the lake for irrigation purposes marred Jackson Hole's natural environment.[13] Was it a mistake under such circumstances, eastsiders might have wondered, for the Reclamation Service to rely on Jackson Hole for water instead of importing it from Yellowstone National Park?

These eastsiders also remained aggrieved enough for a different reason that they would shun any partnerships with federal authorities at Jackson Lake. Eastside irrigationists complained of being the losers when hydrological complications beset the delivery of federally "stored" water down the Snake River channel from Jackson Hole to the Minidoka and Twin Falls tracts. But ambiguity beclouded their claims. When "stored" water emptied from Jackson Lake into the Snake River system, confusion about the property rights to water ensued as a result of "normal flow" water and "stored" water commingling in the Snake River. "Normal flow" water meant those amounts of water that emptied into the Snake under natural conditions—in contrast to spring runoff that the Reclamation Service stored at Jackson Lake and released to irrigationists during the summer. Eastsiders and southsiders at the Twin Falls tracts shared the rights to "normal flow" water. "Stored" water meant water in excess of "normal flow" that the Reclamation Service transferred from Jackson Lake to Minidoka farmers, Twin Falls irrigationists, and certain eastsiders after 1917. However, no hydrological formulations existed for distinguishing between the two classes of water; consequently, there was no way of ensuring that the respective irrigationists received their exact proportions of the "stored" and "normal flow" water that had been intermixed.[14] The Reclamation Service encountered such problematic distinctions between classes of water elsewhere in the West as well.

Apart from quarrels over "normal flow" and "stored" water, eastsiders disputed with southsiders over "seepage" and "return flow" and the place of each in allocating Snake River water. Water seeped from the Snake River during its transit from Jackson Lake, only part of the water reentered the river downstream, and the two sides disputed about this "return flow" into the river. However, hydrological science supplied no means by which the losses and recovery of seepage could be allocated accurately to the eastside and southside groups. Elsewhere in the West, too, the measuring of such water remained an issue. The Idaho sides finally settled at best on "arbitrary agreements" to sort water from year to year. More neighborly in some instances, the groups sacrificed part of their water claims so that none lost all of their crops.[15]

To hold their own, in 1910 eastsiders organized a Farmers Protective Irrigation Association, representing the members of fourteen of the largest irrigation companies and public irrigation districts in the Upper Snake Valley. At first, the association offered theories that, in effect, allocated a larger proportion of all Snake River water to eastsiders. But the association could not enforce its will on the southsiders. Although the latter created no single counterpart to the Protective Association, leaders of the individual Twin Falls projects rejected the new theories; for its part, the

Reclamation Service ignored the Protective Association's ideas and guaranteed Minidoka farmers plenty of water. In the end, the Protective Association retaliated by leading so-called "war[s] of the headgates" against the southsiders in the 1910s.[16] In one case, Aberdeen-Springfield irrigationists intercepted Snake River water before it flowed downstream to southsiders. An observer recounted that in many instances, "armed" eastsiders similarly expropriated water to suit themselves.[17]

Such seizing of water saved eastsiders from losing so many fields of potatoes and sugar beets, both crops that required much watering late in the summer, though the same crops at the Twin Falls tracts suffered from less irrigation as a result of the headgate wars. But the consequences of the wars were more profound. By civil disobedience, eastsiders so hammered their laments into the state's public consciousness that Idaho's civic leaders, state-government officials, and county-level officers were goaded into action. In the years from 1915 to 1919, civic and public leaders offered several proposals by which to lessen eastsider-southsider friction over water and increase the water supply so that farmers could continue to expand row-crop farming in both sectors. Commonly, the new plans entailed storing more of the Snake River system's spring runoff in reservoirs situated within Idaho's borders instead of seeking to expropriate more water and reservoir sites within Jackson Hole. Such planning would avoid new conflicts between Idaho's irrigationists and Wyoming's authorities over Jackson Hole.

Among irrigationists, discussion of the new plans usually revolved around schemes to place new reservoirs at Henrys Lake, at Island Park on the north fork of the Snake River, and at Swan Valley on the south fork of the Snake, and to build a high-rise dam at American Falls. The high-rise dam would obliterate the town of American Falls, waterfalls, and a hydroelectric plant.[18] However, neither eastsiders nor southsiders generally liked the plan in light of what it would cost them to build a new reservoir system at these places. At best, the groups warily compared benefits and prices of such reservoirs. One faction of eastsiders opted out of sharing in the cost of building new irrigation works at American Falls; they feared that southsiders would reduce eastsiders to the status of minority stockholders in the proposed waterworks at American Falls.[19]

When eastsiders and southsiders opposed all of these plans for a new reservoir system, the Reclamation Service proposed its own plan in 1918. Water stored behind a huge dam at American Falls would serve southsiders exclusively, and eastsiders' needs would be satisfied from federally controlled "Jackson Lake [water] storage."[20] This plan displeased both groups of irrigationists. Besides being historically averse to entanglements with federal activities at Jackson Lake, eastsiders also looked askance at the Reclamation Service's plan because Wyoming might someday close off Jackson Hole resources. From the opposite side of the fence, a southsider described the federal plan as unthinkable for his group because it denied southsiders "the desirable qualities of Jackson Lake storage" that they already enjoyed. Among others, the Twin Falls Chamber of Commerce advocated that the southside's irrigationists refuse to pay for any waterworks at American Falls.[21]

These disagreements culminated in an impasse that persisted until the summer of 1919. Finally, because of new hurdles in the way of irrigation farming beginning in this year, irrigationists resumed their dialogue about the old plans of federal, state, and local origin for fixing irrigation in the eastside and southside regions. The new discussions ensued when drought in 1919, dry spells in the next few years, and another drought in 1924 produced water shortages everywhere. Southsiders saw such developments as proof that, contrary to their expectations, water storage at Jackson Lake had given them no security in dry years and, probably could never be expanded enough for political reasons to tide them over dry spells in the future. Finally, many argued that instead of choosing any of the old water plans that were proposed between 1915 and 1919, irrigationists on both sides should solve their problems by importing additional water supplies from Yellowstone National Park. The new proposal relied heavily on findings of the Reclamation Service. From earliest times, the agency deemed it feasible to release water from Yellowstone sources into the Snake River system. Newer findings by engineers also suggested that by so doing, irrigation in the Snake River Basin could be expanded.[22]

In the patchwork of canal companies and irrigation districts in the Upper Snake Valley, leaders of many jurisdictions rallied behind the proposal to import Yellowstone water. Among the first to take such action in 1919, heads of the farmer-owned Fremont-Madison Reservoir Company drew blueprints to extract water from Yellowstone Lake but decided later to draw from the southwestern corner of Yellowstone National Park—the area sometimes identified as Cascade Corner.[23] For the next decade, the group battled with the National Park Service and conservationists for water and two reservoir sites at Cascade Corner.[24] In areas surrounding the Fremont-Madison district, irrigationists schemed to procure water from Yellowstone Lake. Because of their geographic location, many of these groups could readily use such water after it had been emptied from the lake into the south fork of the Snake River; thus utilizing Yellowstone Lake seemed a relatively inexpensive way of augmenting their Snake River water supplies. At the same time, commercial forces at eastside towns such as Rigby, Idaho Falls, and Blackfoot seized on the agitation for Yellowstone water importation to justify their own schemes to resurrect the old Dubois irrigation project. The Reclamation Service had abandoned its plans to reclaim this land in 1906. Nonetheless, these bankers and merchants hoped that with Yellowstone water, several hundred thousand Dubois acres could be turned into an agricultural zone for them to exploit commercially.[25]

Probably Yellowstone Lake could never yield enough water at affordable rates for all of the uses that were proposed: supplement the water supplies of eastsiders and southsiders; provide additional quantities to tide the groups over drought times; and irrigate the expanses of unwatered land in the Dubois project. Satisfying all of these demands for water could even lead to draining away the water resources of Yellowstone, Shoshone, Lewis, and Heart Lakes. Indeed, the new supporters of the Dubois project decided on second thought to rely on water from Shoshone and Lewis Lakes rather than Yellowstone, though Stephen Mather, National Park Service director, denounced their plan.[26]

Enterprisers responded to the demands for export of Yellowstone water to the Snake River Basin, and the best tender came from Charles Carlisle, a civil engineer practicing at Cheyenne, Wyoming. Expecting a profit for himself from diverting water to Idaho, he offered to drain Yellowstone Lake by twenty-nine feet annually. Carlisle believed that by driving a tunnel through the Continental Divide, he could transfer all of this water to the headwaters of the Snake.[27]

A competitor to Carlisle, Idaho enterpriser Ira Perrine also promised to bring water from Yellowstone Lake to the Snake River Basin. Promoting a regional reclamation project of 1,000,000 acres that Perrine named for Secretary of the Interior Franklin Lane, the largest segment of it called Big Bruneau for short, Perrine intended that Yellowstone Lake yield water principally for Big Bruneau's 500,000 acres. The Big Bruneau project stretched westward from the Twin Falls tracts.[28] As in Carlisle's blueprints, Perrine's plan prescribed damming the outlet of Yellowstone Lake so that the water flowed from the lake into cross-mountain tunneling and emptied into the Snake's headwaters. Unlike Carlisle, Perrine sought just the lake's "flood water"—meaning water that would cause flooding in Montana's Yellowstone River Valley were it to flow from the lake in springtime.[29] By one estimate, Perrine's Yellowstone Lake diversion works and his downstream facilities in Idaho would cost $44,000,000.[30]

Implementing his plan over several months in 1919 and 1920, Perrine first persuaded Lane, the irrigation-minded Secretary of the Interior, to allow surveying within the national park. This work proved to Perrine that it was practicable for him to proceed. Next, on the strength of his surveyors' findings, Perrine secured $5,000,000 in seed money from private sources; reportedly, Idaho Power Company, the Electric Bond and Share Company of New York City, and the Union Pacific Railroad bankrolled him. All expected to benefit financially from capitalizing Perrine's share of developing Idaho's irrigation frontier. Perrine earmarked their funds for transmontane tunneling to link Yellowstone Lake and the Snake River watershed.[31]

At the same time, influential forces boosted for Perrine's and Carlisle's plans. Idaho governor David Davis defended the schemes.

Wyoming governor Robert Carey stated that water from the lake was best utilized for wet farming in Idaho and Montana inasmuch as it "cannot be applied to irrigation in Wyoming."[32] Early in 1920, Secretary Lane renewed his backing for Perrine's scheme.[33]

Replying to the irrigationists in 1919 and 1920, the National Park Service and its conservationist allies mounted their first concerted attack on the schemes of the Fremont-Madison forces, Carlisle, and Perrine. Objecting on conservationist grounds to any agricultural intrusions at Yellowstone National Park, these critics also charged that Carlisle and Perrine would ruin Yellowstone Lake and its surroundings environmentally but run out of enough capital to succeed. Perrine partly rebutted the last charge by trying to capitalize his Big Bruneau project differently. He demanded Idaho's share of $100,000,000 that Secretary Lane wanted to spend on resettling World War I veterans on new farmsteads. Carlisle eyed the same money. Perrine and Carlisle also hoped that Congress would respond favorably to proposals from the Western States Reclamation Association, a body that Governor Davis organized in 1919. The association asked for federal spending of another $250,000,000 on arid land reclamation.[34]

Congress appropriated no money for either of these federal programs; nonetheless, Perrine and Carlisle continued to eye Yellowstone Lake and calculate their chances of paying for the diversion of its water entirely out of private funds. Finally, in 1920, their persistence so alarmed Mather that the National Park Service director struck back at them in another way. He invoked his powers in 39 Stat. 535 (1916), a federal law creating the National Park Service, so that Perrine and Carlisle could not "injure the scenic beauty" of Yellowstone Lake and its environs, intrude on any pristine wilderness to construct their tunnels, or "play havoc" at Heart Lake. They had planned to situate their last catchment there before transferring Yellowstone Lake water to the Snake River Basin.[35]

Mather also invoked his statutory powers to protect the natural environment around Shoshone and Lewis Lakes. It bothered him that the newest promoters of Dubois irrigation coveted water from those lakes for irrigation of their project. In a worst outcome, Mather speculated, Carlisle and Perrine would as a last resort take all of their booty at Shoshone and Lewis Lakes in order to appease certain Montanans who had always wanted to monopolize Yellowstone Lake for their own purposes. Mather believed it likely that Perrine, Carlisle, and the Montanans could persuade Congress to divide Yellowstone National Park waters in ways that the irrigationists asked.[36]

Continuing his fight, Mather secured the backing of powerful groups, among them the American Civic Association, American Automobile Association, and National Parks Association. This last group accused irrigationists of waging "War on the National Parks."[37] Followers of naturalist John Muir, different activists whom scholars sometimes characterize as preservationists, supported Mather. They held that Yellowstone National Park should be closed to agriculturalists in light of what had happened when commercialism had intruded a decade earlier at Hetch Hetchy Valley in Yosemite National Park. Preservationists helped to organize a broader coalition committed to Yellowstone "park defense." Among others, the new federation contained wilderness and wildlife savers, women's professional and social organizations, scientific societies, commercial groups, and fine arts associations.[38]

Mather next escalated the water war at the irrigationists' expense. First, he and his supporters persuaded Congress to withhold Yellowstone water from Perrine, Carlisle, and the Fremont-Madison forces and then ridiculed the irrigationists' high-minded rebuttals against the conservationists. In the end, the irrigationists could not prove that public interests would be furthered if Yellowstone water sustained irrigable terrain. Also to no avail, Perrine claimed that were Yellowstone Lake's "flood water" transferred to Idaho, it would stop the water from annually inundating low lands in Montana and, on proceeding downstream, worsening flooding in the Missouri-Mississippi Valley.[39]

In 1920, Idaho irrigationists also lost ground after their best ally in federal circles, Secretary Lane, resigned from the government. Subsequently, Mather and his supporters converted John Payne, Lane's successor, to their cause; Secretary Payne then ordered that agriculturalists not intrude at Yellowstone National Park "under any circumstances."[40] To no avail, Perrine and Carlisle challenged Payne's ruling. For their part, Fremont-Madison forces petitioned Congress to transfer Cascade Corner from the National Park Service to an irrigation-friendly federal agency. Payne and Mather blocked this maneuver.[41]

At the same time, Perrine and Carlisle failed to counter new danger to their schemes from the Yellowstone Irrigation Association, a mouthpiece for irrigationists in the Yellowstone River Valley. The association, claiming for Montanans all "flood water" from Yellowstone Lake, attempted to enforce such a monopoly.[42] At last, the Montanans prevailed with help from outsiders. In Wyoming, Governor Carey endorsed their irrigation plans. U.S. Reclamation Service engineers decided that the Montanans "presumably" held better claims than Idahoans to Yellowstone Lake water. Given this conclusion, Arthur Davis, director of the Reclamation Service, added that Idahoans deserved no Yellowstone water "until the[ir] need for it became more pressing."[43]

Thwarted at every turn, Perrine and Carlisle still persisted. However, federal authorities blocked them from ever bringing Yellowstone water to the Snake River Basin. In 1920, Payne continued to enforce his closure of Yellowstone National Park to agriculturalists; President Warren Harding permitted the same closure policy to stand from 1921 to 1923; and President Calvin Coolidge followed suit after 1923.[44]

———◆———

Denied Yellowstone water during extraordinarily dry times, eastsiders and southsiders reviewed their water situation and concluded from this rethinking—a process made more harried by new drought in 1924—that both camps must somehow acquire what was termed "supplemental" water supplies. Accordingly, southsiders relented out of adversity to what they had opposed in earlier days—damming the Snake River at the American Falls. Soon southsiders were even at the forefront of a movement to adopt the old strategy of capturing much of the spring runoff into the river by putting an expensive high-rise dam at American Falls. Leaders of a few eastside irrigation organizations agreed to participate in such dam building. The forces created their own irrigation district. Finally, the new district sold its bonds to commercial investors and raised millions of dollars with which to reimburse the U.S. Bureau of Reclamation (successor to the Reclamation Service after 1923) for constructing the new dam. On line by 1927, the facility could impound 1,700,000 acre-feet of water.[45]

Eastsiders mostly dithered about how to acquire enough "supplemental" water to remedy their water shortages. Throughout the 1920s, many continued to reject damming at American Falls as a substitute for importing Yellowstone Lake water even though the latter choice had become a lost cause. Fremont-Madison irrigationists went their own way, after each setback renewing the group's old fight for water and reservoir sites at Cascade Corner. This water war lasted until 1930. Notably in Bonneville and Jefferson Counties, different groups of eastsiders argued inconclusively about several proposals to collect "supplemental" water at sites along the Snake River's north and south forks and build reservoirs nearby in the Teton River Basin. By one account, utilizing the best of such reservoir sites, especially ones along the water-rich reaches of the south fork, was recommended despite the high price of building a new reservoir system in the eastsiders' backyards. Such a system, analysts contended, would ensure better irrigation at the old wet-farm tracts, banish aridity at last from the long-defunct Dubois project of 200,000 acres, and allow watering of another 400,000 acres of "new land" scattered around the Upper Snake Valley's several counties.[46]

Furthermore, certain eastsiders argued that although Wyoming and United States officials and conservationists had forever closed off any greater access to Jackson Lake, eastsiders might be able to exploit the remainder of Jackson Hole's water riches. For example, smaller lakes could be accessed by buying the decreed water rights of Wyomingites at these points. Wyoming entrepreneurs seized on the proposition. Offering to round up and sell water claims at such places, the enterprisers particularly touted a proposal to dump water from Jenny, Leigh, Emma Matilda, and Two Ocean Lakes into the Snake River system. Idaho state government leaders also endorsed the scheme but could not convince Wyoming authorities to authorize such diverting of water. In one controversy, an Idaho official attributed the outcome to "unusual conservatism, bordering on selfishness" in state-level circles at Cheyenne.[47] Mather and his conservationist allies also objected to this scheme because they hoped eventually to enfold Jackson Hole into the national park system: the National Park Service had begun a new fight that culminated in the creation of Grand Teton National Park in 1929.[48]

William Cox offered eastsiders a new proposition in 1924. Head of the Cheyenne-based Teton Irrigation Company, which owned many filings on Jackson Hole water, Cox was prepared to supply water to eastsiders from Jenny and Leigh Lakes, the Gros Ventre River, and two smaller streams.[49] Cox could not succeed. The National Park Service and its allies objected on grounds that no irrigation works belonged anywhere in Jackson Hole. At first, Wyoming authorities barred Cox from tampering with Jenny and Leigh Lakes, both of which they deemed "real gems from a scenic viewpoint."[50] Subsequently, the same officials scuttled all of Cox's plan. Listening to groups who demanded that Jackson Hole become a recreation zone, they decided never to tolerate new "eyesores" like the one irrigationists had already built at Jackson Lake.[51]

When "supplemental" water could not be procured at Jackson Hole, eastsiders disagreed again about ways to eliminate their water shortages. Beginning in 1924, they stored extra water at Henrys Lake, a site near the head of the Snake River's north fork [also called Henrys Fork]; but that action fixed problems only for the handful of eastsiders participating in dam building at American Falls. Then water shortages mounted across the Upper Snake Valley in the drier years from 1926 to 1932. Compounding these troubles, the Bureau of Reclamation promised but reneged on locating sites along the Snake River's north and south forks where water could be collected economically.[52] Next, drought worsened in 1933 and 1934, the latter the driest year for Idaho since weather recording began in 1909. By one estimate, water shortages caused losses of eight to ten million dollars on about 1,000,000 eastside acres.[53] In the drought of 1934, southsiders also lost many crops because too little water had accumulated for them in reservoirs at Jackson Lake and American Falls. Eastsiders also seized part of their scarce supply of water.[54]

For Snake River irrigationists, the latest drought proved again that all were vulnerable to water shortages in the worst of times. It was clear that even those groups holding the best water-storage rights at Jackson Lake and American Falls needed somehow to keep greater surpluses of water on hand in order to avoid new disasters comparable to what happened in 1933 and 1934.

Many eastsiders responded to the new crisis by calling again for Yellowstone Lake water, and even the Fremont-Madison forces supported this demand after being advised that "scenic apostles" would always block them from procuring their supplemental water at Cascade Corner.[55] Partly these eastsiders echoed old notions that it was less expensive to import water from the outside than to build new reservoirs in the Upper Snake Valley. But they also rationalized their call for Yellowstone Lake water on grounds that hopes of acquiring it were no longer a pipedream because of government opposition. With new leadership at the National Park Service and the assumption of power nationally by Franklin Roosevelt and his New Dealers in 1933, the irrigationists theorized, such federal forces were likely to confront the Great Depression of the 1930s with measures that even defied conservationist traditions. Moreover, government-based reclamationists egged on the eastsiders. One official proposed that they annually exploit 500,000 acre-feet of water from Yellowstone Lake. Robert Faris, Idaho' s commissioner of reclamation, argued for commandeering 300,000 acrefeet of water from the same lake. Idaho governor C. Ben Ross endorsed the Faris plan in 1933.[56]

The Faris Plan, copying old schemes formulated in the 1920s, entailed blocking Yellowstone Lake's outlet to collect "flood water" for irrigation.[57] Such a procedure, Faris argued, risked nothing environmentally at the national park because an annual draw of 300,000 acre-feet from the lake would lower it by five feet—well within the lake's natural fluctuation from high to low water mark. Faris' plan also provided for protecting the park's scenery, wilderness, and recreational values. The plan required that water be diverted at a remote arm of Yellowstone Lake and pass from there through tunnels, hidden in mountains and virtually inaccessible terrain, to reach tributaries of the Snake River. Building this water-diversion system in Great Depression times probably would cost $6,750,000.[58]

After treating first with Arno Cammerer, director of the National Park Service, who rejected the plan, Ross presented the same plan to Roosevelt and his New Dealers. But Ross entreated to no avail. In 1933, Secretary of the Interior Harold Ickes declared Yellowstone Lake off limits to Idaho's farm interests, a ban that he applied simultaneously to Montana forces that also wanted to use the lake for irrigation purposes.[59] Roosevelt upheld Ickes. Privately, he also reassured conservationists that he would fend off irrigationists at Yellowstone National Park; publicly, Roosevelt reaffirmed this stance during official appearances at the park in 1934 and 1937.[60]

Ross prodded Cammerer to revisit the Faris Plan, and the next state governor, Barzilla Clark, renewed the pressure on Cammerer to reconsider in 1937. Cammerer refused. He wrote that opposing the Faris Plan was an "absolute necessity" from conservationist perspectives.[61] Cammerer also resisted for other reasons. He suspected eastsiders of harboring designs on more than Yellowstone Lake in the absence of new prerogatives for themselves at Jackson Hole. Additionally, Cammerer was alarmed when eastside commercial elements claimed again that the defunct Dubois project could eventually receive water from Yellowstone Lake. Such ambitions suggested to him that eastsiders actually coveted the water resources of Yellowstone, Shoshone, Lewis, and Heart Lakes.

Cammerer had qualms, too, lest southside groups petition for water from several of the same lakes in Yellowstone National Park in order to effect first-time irrigation at Big Bruneau. Ira Perrine, who had proposed irrigation development there in 1919, revived the Big Bruneau project in 1934. He judged it viable and expected to tap California sources of capital for its development despite criticism that Big Bruneau was "a wild scheme which could never be built."[62]

Ignoring Cammerer's stonewalling, eastsiders again petitioned Department of the Interior officials for water from Yellowstone Lake. In part, they acted hastily because commercial and farm interests in Wyoming and Montana were now competing with them for such water.[63] The eastsiders' drive gathered momentum in 1937. The Idaho legislature helped them when, on February 4, the body adopted a memorial to Congress on the eastsiders' behalf.[64] Furthermore, Faris speculated that eventually New Dealers would accept his water plan. As the thirties had passed, he argued, New Dealers had exhibited much inventiveness and liking for social and economic experimentation. By capitalizing on this new climate, where "stranger things have [already] happened," Faris theorized that eastsiders could finally extract permission from New Dealers to take Yellowstone water.[65] It also delighted Faris that, in 1937, New Dealers had created a new precedent by allowing irrigationists to convey water across Rocky Mountain National Park for use on the Big Thompson irrigation project in Colorado.[66] Faris invoked that development as precedent for irrigationists to exploit Yellowstone water resources. New Dealers ignored him.

Trying another tack, Faris attempted to convince New Dealers that his plan would lessen flooding in the nation's Mississippi-Missouri heartland. By Faris' arguments to them, impounding "flood water" from Yellowstone Lake for diversion to Idaho in the summer would prevent this water from causing floods in Montana's Yellowstone River Valley on emptying from the lake, and from flowing downstream to contribute to new flooding in the Missouri and Mississippi Valleys. Hydrologists at the Idaho Department of Reclamation confirmed Faris' assumptions.[67] However, these arguments never interested New Dealers because Faris' theory merely rehashed old ideas that conservationists had rebutted in the 1920s. New Dealers also cited new hydrological data to refute Faris.[68]

Although nothing ever budged New Dealers from barring irrigationists from Yellowstone water and reservoir sites, Roosevelt thought as early as 1934 that New Dealers should respond in other ways to Idaho's irrigationists. He believed that they had made a persuasive case for obtaining more water.[69] Accordingly, Secretary Ickes approved several initiatives for helping eastside and southside irrigationists with federal dollars. Offering federal financing of new irrigation works outside Yellowstone National Park, he allocated most of this money to Idaho developments from large sums that Congress had appropriated to Ickes' Public Works Administration.[70] In this way, he extended the federal government's long-term commitments to promoting irrigation and reclamation in Idaho.

In one case, Ickes allocated $4,000,000 so that Fremont-Madison irrigationists could procure extra water within the basin of the Snake River's north fork—provided the group abandon all suggestions of Yellowstone water-taking. The Fremont forces hesitated but finally reached a meeting of the minds with Ickes. On his terms, United States money eventually capitalized north-fork waterworks at Island Park and a smaller facility at Grassy Lake, a site close to the Idaho-Wyoming border. By the end of the 1930s, the new works annually impounded about 144,000 acre-feet of spring runoff water.[71]

To procure even greater relief for irrigationists, Ickes decided to spend federal dollars on commandeering water along the water-rich stretches of the Snake River's south fork. Hoarding the fork's runoff in wet years, he believed, would in the future tide eastsiders and southsiders over cycles of dry years. Ickes' plan conflicted with the disposition of his own Bureau of Reclamation to rely on Yellowstone and Jackson Hole sources for new water supplies. But the Bureau, choosing not to clash with Ickes, cooperated with him by mounting new searches for acceptable dam and reservoir sites along the south fork.[72] Bureau officials also ruled—without challenge from Ickes—that unirrigated tracts like Dubois and Big Bruneau were ineligible for help because south-fork water resources belonged on land "already settled and improved."[73]

Governor Ross encouraged such usage of south fork water although he continued to tout the Faris Plan for tapping Yellowstone Lake water, and in 1935 he proposed that Ickes spend $6,500,000 on south-fork dam and reservoir developments.[74] On succeeding Ross in 1937, Barzilla Clark adopted a two-pronged policy: water-works development on the south fork and water importation from Yellowstone Lake.[75]

For the most part, southsiders subscribed to Ickes' scheme for creating grand irrigation works on the Snake River's south fork. Because they would receive part of the new water supply, they approved the federal plan even though the same southsiders preferred that more of the Snake River's runoff be stored at the American Falls Reservoir.[76] Southsiders liked Ickes' plan even better when they learned that the Bureau of Reclamation favored the development of a grand damsite on the south fork known as the Palisades. The site was ideal by engineering and economic criteria save one—a dam there would cost slightly more than similar works at Johnny Counts, a site across the border in Wyoming. But damming at the Palisades avoided legal and political conflicts with Wyoming over Johnny Counts; better yet, Wyoming and Jackson Hole authorities could exercise little power at the Palisades because the dam and nearly all of the reservoir behind it would be within Idaho.[77]

In 1938, several groups of eastsiders mounted a last-ditch fight for Yellowstone Lake water in place of relying on the Snake River's south fork to yield more irrigation waters. Observers accused the groups of harboring a pipe-dream, but the holdouts reiterated their traditional notions that importing water was the cheapest way of augmenting the Upper Snake Valley's supply. Besides hoping to pinch pennies in this way, many holdouts dreamed of prevailing when Governor Clark and Idaho's Democratic Party organizations would force New Dealers to relent from protecting Yellowstone National Park. Holdouts had, however, exaggerated the influence of Idaho's Democrats in New Deal circles. James Pope, Idaho's lone Democrat in the United States Senate, offered legislation that allowed the holdouts to secure water from Yellowstone Lake. The new bill never passed. Pope blamed Ickes, accusing the Secretary of the Interior of pacifying the National Park Service and bullying the Bureau of Reclamation into opposing Pope's legislation. Pointing to different bogeymen, an irrigation-district official attributed the outcome to blocking by "nature lovers" and agricultural and commercial forces in Montana and Wyoming. The latter, he believed, had drubbed eastsiders in order to protect their own interests at Yellowstone Lake.[78] In any event, Faris added, the drive to import Yellowstone water to the Snake River Basin had devolved into "a chimerical dream comparable to a child crying for the moon."[79]

Beaten again in the long fight for access to Yellowstone Lake, eastsiders mostly agreed with Faris that probably their prospects for securing Yellowstone Lake water were trashed permanently. Speaking for numerous irrigation organizations in 1939, the Upper Snake River Valley Water Users' Protective Union endorsed the federal scheme for damming at the Palisades in lieu of importing Yellowstone water. In fact, an observer claimed, eastsiders everywhere "have waked up" to acknowledge that obtaining water storage at the Palisades beat fighting for Yellowstone Lake.[80] Finally, in 1957, a high-rise dam at the Palisades came on line. Capable of impounding 1,400,000 acre-feet of runoff water, the new waterworks greatly augmented the water resources available to eastsiders and provided greater security for residents on the Twin Falls and Minidoka tracts.[81]

For different reasons, too, old impulses diminished either to import Yellowstone water or to expropriate more water resources at Jackson Hole. Water shortfalls eased for most of southeastern Idaho's irrigation tracts in the wetter years of the forties and fifties, leaving the eastside/southside camps notably more contented with the New Dealers' water decisions of the 1930s. In an interstate water compact of 1949, Idaho and Wyoming discouraged eastsiders and southsiders from ever again demanding more water and reservoir sites at Jackson Hole.[82]

Of the 3,000,000-acre irrigation frontier in the eastside/southside half of the Snake River Basin, about 60 percent had been reclaimed by the middle of the twentieth century. Great expanses of land at Dubois, Big Bruneau, and other irrigable stretches waited only for water to transform the land into agriculturally productive terrain. But the grander extent of successful land reclamation in southeastern Idaho in just fifty years represented victories for irrigationists who had persevered through essentially the same contentious discourses on water usage that characterized western dialogues everywhere about irrigation and arid-land reclamation. In southeastern Idaho, as in much of the West's irrigation politics, the issues in such political decision-making had ranged over local, state, and federal irrigation policies, federal and state powers over irrigation and land reclamation, interstate water jurisdictions, and the control of natural resources.

In reaching such a point in the conquest of the eastside/southside irrigation frontier, irrigationists lost their battles to divert water into the Snake River Basin from Yellowstone Lake, to use Cascade Corner for irrigation purposes, and to exploit Jackson Hole's water resources except at Jackson Lake. Bureaucratic infighting at the federal level, interstate rivalry over the Snake River system's water, and environmental activism complicated this struggle—and helped to shape the outcome of the fighting. However, the fight also lengthened because the last holdouts persevered; they finally abandoned their push for Yellowstone Lake water diversion only long after the federal government opposed such a move starting early in the 1920s. In the end, the federal government played a crucial role in the interplay of forces engaged in such irrigation politics over Yellowstone and Jackson Hole water, and the federal solution of building grand reservoirs at American Falls, Island Park, and the Palisades was finally forced on the recalcitrant irrigationists of southeastern Idaho.

More important, the conquest of so much of Idaho's eastside/southside irrigation frontier in spite of irrigationists being excluded wholly from Yellowstone National Park and partly from Jackson Hole occurred because of the federal government's long-term commitment to expanding irrigation and land reclamation in Idaho. Beginning with the Carey Act of 1894, under which reclamation became more successful in Idaho than in the remainder of the West, and later under the Reclamation Act of 1902, Idaho long benefitted from the United States' commitment to "making the desert bloom." Over time, the state got millions of dollars in federal aid and two major federal water projects. Among such benefits, the federal government played a big role in advancing irrigation development in the eastern half of the Snake River Basin. In 1940, a federal reclamationist bragged that besides providing "government" water to the Minidoka and Boise projects, the Bureau of Reclamation already furnished supplemental water supplies to 800,000 acres. The latter were situated mainly in southeastern Idaho. By another set of measurements made after the federally constructed waterworks at the Palisades were finished in 1957, 1,250,000 acres in the eastside and southside sectors received 8,200,000 acre feet of water in a single year partly because of federal contributions to bettering Snake River Basin irrigation.[83] Hence the fate of Idaho's eastside/southside irrigation frontier and federal water policies had become even more entwined.

This article originally appeared in *Idaho Yesterdays* 43, no. 4 (Winter 2000). Reprinted with permission.

Notes

1. *Wendell (ID) Irrigationist*, March 4, 1915; Clyde H. Porter to Moses Alexander, September 15, 1915, Moses Alexander Papers, Idaho State Archives, Boise (hereafter ISA) (qtn.).

2. Elwood Mead, *Irrigation Institutions: A Discussion of the Economic and Legal Questions Created by the Growth of Irrigated Agriculture in the West* (1903; reprinted, New York: Johnson Reprint Corporation, 1972), 353; J. W. Powell, *Eleventh Annual Report of the United States Geological Survey to the Secretary of the Interior, 1889–90, Part II–Irrigation* (Washington, DC: Government Printing Office, 1891), 191.

3. "History of the Minidoka Project, Idaho to 1912," typescript, Volume 1, 5–6, Project Histories and Reports of Reclamation Bureau Projects: Minidoka Project, Idaho, 1910–1919 (hereafter Minidoka Project), U.S. Bureau of Reclamation Records, Record Group 115 (hereafter USBRR), National Archives and Records Administration, Microfilm Roll 96; *The Twin Falls Canal Company System* (Twin Falls, ID: Twin Falls Canal Company, 1929), 3, 4.

4. On the two federal projects in Idaho, see *Project Data 1981*, Water and Power Resources Service, U.S. Department of the Interior (Denver: Government Printing Office, 1981), 43–60, 639–654. New scholarship of Norris Hundley, Donald Worster, and Donald Pisani has greatly increased and improved the historical literature on federal reclamation work; for example, see Donald J. Pisani, *To Reclaim a Divided West: Water, Law, and Public Policy, 1848–1902* (Albuquerque: University of New Mexico Press, 1992), 273–319. For samples of new studies of single projects, see Karen L. Smith, *The Magnificent Experiment: Building the Salt River Reclamation Project, 1890–1917* (Tucson: University of Arizona Press, 1987), and Paul C. Pitzer, *Grand Coulee: Harnessing a Dream* (Pullman: Washington State University Press, 1994). On the Snake River experience as a whole, see Mark Fiege, *Irrigated Eden: The Making of an Agricultural Landscape in the American West* (Seattle: University of Washington Press, 1999).

5. [Fred Dubois] to Secretary of the Interior, April 11, 1904, Fred T. Dubois Papers, Special Collections, Idaho State University Library, Pocatello (hereafter SC-ISU); Patricia Lyn

Scott, "Dubois: The Project that Never Was," *Snake River Echoes* (1991), 20:38–40.

6. *Second Annual Report of the Reclamation Service, 1902–03*, U.S. Department of the Interior, U.S. Geological Survey, House Doc. 44, 58th Cong., 2d sess. (Washington, DC: Government Printing Office, 1905), 276 (qtn.); J. G. Camp, "Parks—and Homes in [the] Snake River Valley," *New West Magazine* (December, 1920), 11:43.

7. *Second Annual Report*, 272–274; "Notes for Senator Dubois," June 15, 1905, File 123, Box 32/917128, USBRR, Federal Archives and Records Center, Denver. All USBRR material cited in manuscript form is at the Denver FARC; other USBRR material is identified here by microfilm roll number.

8. *Second Annual Report*, 275–276, 277.

9. *Twin Falls Canal*, 21; Hugh T. Lovin, "Free Enterprise and Large-Scale Reclamation at the Twin Falls-North Side Tract, 1907–1930," *Idaho Yesterdays* (Spring, 1985), 29/1:11–13 [Chapter 5 in this book]; S. O. Harper, "39 Years of Federal Reclamation in Idaho," typescript, n.d. [May 10, 1941], 3, Sinclair O. Harper Papers, American Heritage Center, University of Wyoming, Laramie (hereafter AHC). For accounts of Carey Act reclamation in Idaho, see Hugh T. Lovin, "The Carey Act in Idaho, 1895–1925: An Experiment in Free Enterprise Reclamation," *Pacific Northwest Quarterly* (October, 1987), 78:122–133 [Chapter 4 in this book], and Mikel H. Williams, *The History of Development and Current Status of the Carey Act in Idaho* (Boise: Idaho Department of Reclamation, 1970).

10. "Jackson Lake Dam the Savior of the Snake River Valley," *Engineering News-Record* (December 11–18, 1919), 83:992–994.

11. Fiege, *Irrigated Eden*, 103.

12. Leonard J. Arrington, *History of Idaho*, 2 volumes (Moscow: University of Idaho Press, 1994), 2:158; *Ninth Biennial Report of the State Engineer to the Governor of Idaho, 1911–1912* (N.p: n.p., 1912), 337–338.

13. T. A. Larson, *History of Wyoming* (Lincoln: University of Nebraska Press, 1965), 358, 420.

14. "Minidoka Project History, 1916," typescript, 132–143, USBRR, Roll 99; Fiege, *Irrigated Eden*, 99–102.

15. "Report of [the] Sixth Annual Conference of Operating Engineers Held in Boise, Idaho, January 15–18, 1917," mimeographed, 95–96, Project Histories and Reports of Reclamation Bureau Projects: Boise Project, Idaho, 1906–1921, USBRR, Roll 23.

16. Kate B. Carter, *Pioneer Irrigation, Upper Snake River Valley* (N.p.: Daughters of Utah Pioneers, 1955), 294.

17. "Minidoka Project History, 1915," 80–81, USBRR, Roll 98; A. E. Robinson to Moses Alexander, October 10, 1914, Alexander Papers (qtn.).

18. F. A. Wilkie to Lorenzo R. Thomas, January 22, 1919, Idaho Reclamation Records (hereafter IRR), ISA; F. A. Banks to Chief of Construction, September 18, 1919, Engineering Correspondence Files, Box 787, USBRR.

19. South Fork of the Snake River Reservoir Association, "Ten Reasons," typescript, n.d., Engineering Correspondence Files, Box 787, USBRR; *Idaho Register* (Idaho Falls), August 5, 1919.

20. *Eighteenth Annual Report of the Reclamation Service, 1918–1919* (Washington, DC: Government Printing Office, 1919), 414.

21. *Wendell Irrigationist*, June 22, 1922 (qtn.); Irvin E. Rockwell, *The Saga of American Falls Dam* (New York: Hobson Book Press, 1947), 16–17.

22. "Report of the Board of Engineers to Consider Projects in [the] Snake River Valley...," mimeographed, April 10, 1920, iv, 8, 40–41, Ernest G. Eagleson Papers, Idaho State Historical Society, Boise (hereafter ISHS).

23. John J. Cameron, "Proposed Irrigation Projects within the Yellowstone National Park," typescript (N.p.: U.S. Department of the Interior, National Park Service, May, 1937), 15, 16 (copy in the Yellowstone National Park Research Library, Yellowstone National Park, Wyoming); *Ashton (ID) Herald*, November 11, 1920.

24. Richard A. Bartlett, *Yellowstone: A Wilderness Besieged* (Tucson: University of Arizona Press, 1985), 355–357.

25. W. G. Swendsen to D. W. Davis, December 29, 1919, IRR; William Darrell Gertsch, "The Upper Snake River Project: A Historical Study of Reclamation and Regional Development, 1890–1930" (PhD diss., University of Washington, Seattle, 1974); *Rigby (ID) Star*, September 25, 1919, January 15, 1920; H. V. Carleton, "The Great Dubois Project in Idaho," *New West Magazine* (June–July, 1919), 10:8–10.

26. Cameron, "Irrigation," 16; *Report of the Director of the National Park Service to the Secretary of the Interior for the Fiscal Year ended June 30, 1920 and the Travel Season 1920* (Washington, DC: Government Printing Office, 1920), 23–25.

27. Cameron, "Irrigation," 16; Horace M. Albright (as told to Robert Calm), *The Birth of the National Park Service: The Founding Years, 1913–33* (Salt Lake City: Howe Brothers, 1985), 102.

28. "Franklin K. Lane Reclamation Project [of] One Million Acres in Southern Idaho, According to Plans Developed by I. B. Perrine and H. L Hollister," n.d., William E. Borah Papers, Library of Congress, Washington, DC.

29. W. G. Swendsen, "Preliminary Report on the Possibilities of Diverting Yellowstone Lake [Water] into the Snake River Drainage," typescript, n.d. [1919], no pp., IRR.

30. Paul S. A. Bickel, "Report on [the] Bruneau Irrigation Project," typescript, April 9, 1920, 10–11, File 125–1, Box 1/SB203132, USBRR.

31. Ernest G. Eagleson to Franklin K. Lane, July 26, October 30, 1919, and Eagleson to Charles L. Hyde, January 19, 1920, Eagleson Papers.

32. W. G. Swendsen to R. N. Tiffany, December 10, 1920, IRR; *Cheyenne (WY) State Leader*, September 28, 1919 (qtn.).

33. Donald C. Swain, *Wilderness Defender: Horace M. Albright and Conservation* (Chicago: University of Chicago Press, 1970), 122.

34. A. C. Milner to J. W. Sweeley, January 12, 1919 (copy), File 125–1, Box 11SB203132, USBRR; Eagleson to Lane, October 30, 1919; "Chances for Success of $250,000,000 Reclamation Bond Issue," *Engineering News-Record* (February 12, 1920), 84:347.

35. Robert Shankland, *Steve Mather of the National Park Service*, 2nd edition (New York: Alfred A. Knopf, 1954), 212–213; *Report of the Director of the National Park Service...1920*, 21–30 (qtns. on 22, 23).

36. *Report of the Director of the National Park Service...1920*, 23–24; Cameron, "Irrigation," 16.

37. Ernest Morrison, *J. Horace McFarland: A Thorn for Beauty* (Harrisburg: Pennsylvania Historical and Museum Commission, 1995), 236–239; John C. Miles, *Guardians of the Parks: A History of the National Parks and Conservation Association* (Washington, DC: Taylor and Francis, 1995), 33 (qtn).

38. "Another Hetch Hetchy," *The Outlook* (July 7, 1920), 125:48; Alfred Runte, *National Parks: The American Experience*, 2nd rev. edition (Lincoln: University of Nebraska Press, 1987), 108; W. C. Gregg, "Cascade Corner of Yellowstone Park," *The Outlook* (November 21, 1921), 129:476 (qtn).

39. For Perrine's position on flooding, see W. G. Swendsen to R. N. Tiffany, December 13, 1920, IRR.

40. Albright, *Birth*, 107.

41. Aubrey L. Haines, *The Yellowstone Story: A History of Our First National Park*, 2 volumes (Yellowstone Park, Wyoming: Yellowstone Library and Museum Association/Colorado Associated University Press, 1977), 2:330–334.

42. *Billings (MT) Gazette*, December 7, 1919; *M. M. Galbraith and others, Report on [the] Proposed Project for Flood Control and Irrigation in the Yellowstone River Valley...* ([Livingston, Montana: Yellowstone Irrigation Association], 1921), copy at Montana Historical Society, Helena (hereafter MHS).

43. "Report of the Board of Engineers...," 40 (1st. qtn.); A. P. Davis to Chief Engineer, June 1, 1920, Engineering Correspondence Files, Box 787, USBRR (2nd qtn.).

44. Shankland, *Mather*, 85–86; John Ise, *Our National Park Policy: A Critical History* (Baltimore: Resources for the Future by The Johns Hopkins University Press, 1961), 314.

45. Gertsch, "Upper Snake," 2325–341.

46. Rockwell, *Saga*, 104–107; *Second Biennial Report of the Department of Reclamation, State of Idaho, 1921–1922* (N.P.: n.p., 1922), 250 (qtn.).

47. Robert W. Righter, *Crucible for Conservation: The Creation of Grand Teton National Park* (Boulder: Colorado Associated University Press, 1982), 32, 33; W. G. Swendsen to D. W. Davis, January 22, 1921, D. W. Davis Papers, ISA (qtn.).

48. *Report of the Director of the National Park Service to the Secretary of the Interior for the Fiscal Year Ended June 30, 1923 and the Travel Season 1923* (Washington, DC: Government Printing Office, 1923), 51; Horace M. Albright to William B. Ross, January 26, 1923, William B. Ross Papers, Wyoming State Archives, Cheyenne (hereafter WSA); Righter, *Crucible*, 43ff.

49. William F. Cox to Charles C. Carlisle, April 13, 1924, January 28, March 7, 1925, Charles C. Carlisle Papers, WSA.

50. *Seventeenth Biennial Report of the State Engineer to the Governor of Wyoming, 1923–1924* (Cheyenne: n.p., 1924), 24.

51. Ibid., 23–24; Robert H. Betts, *Along the Ramparts of the Tetons: The Saga of Jackson Hole, Wyoming* (Niwot: University Press of Colorado, 1978), 196–97; David J. Saylor, *Jackson Hole, Wyoming: In the Shadow of the Tetons* (Norman: University of Oklahoma Press, 1970), 155 (qtn).

52. C. Ben Ross to Horace M. Albright, August 11, 1933, C. Ben Ross Papers, ISA; *Eighth Biennial Report of the Department of Reclamation, State of Idaho* (N.p.: n.p., 1934), 7.

53. Carlos Arnaldo Schwantes, *The Pacific Northwest: An Interpretive History*, rev. edition (Lincoln: University of Nebraska Press, 1996), 383; R. W. Faris to Harold L. Ickes, September 11, 1934, Ross Papers.

54. Arrington, *History*, 2:28–29.

55. *Idaho Falls Post-Register*, November 16, 1934; William E. Borah to R. W. Faris, June 28, 1937, IRR (qtn.).

56. R. W. Faris, "Supplementary Water for Irrigation in Idaho, with Particular Reference to the Boise and Snake River Valleys," typescript, n.d., 12, and R. W. Faris to J. P. Pope, July 23, 1933, IRR; C. Ben Ross to Horace M. Albright, August 11, 30, 1933, Ross Papers, ISA.

57. Faris, "Supplementary Water," passim.

58. Cameron, "Irrigation," 42a–42d; R. W. Faris to Arno D. Cammerer, May 25, 1937, IRR; *Ninth Biennial Report of the Department of Reclamation, State of Idaho, 1935–1936* (N.p.: n.p., 1936), 26.

59. Compton I. White to R. W. Faris, July 24, 1933, and J. P. Pope to Faris, August 1, 1933, IRR; Harold L. Ickes to Frank H. Cooney, May 12, 1933, Frank H. Cooney Papers, MHS.

60. Excerpts from Franklin Roosevelt's public comments, August 5, 1934, and Roosevelt to Louis B. DeKoven, June 8, 1938, in Edgar B. Nixon, editor, *Franklin D. Roosevelt and Conservation, 1911–1945*, 2 volumes (1957; reprint, New York: Arno Press, 1972), 1:323, 2:234; Bartlett, *Yellowstone*, 358.

61. Arno B. Cammerer to R. W. Faris, June 14, 1937, IRR.

62. "Idaho Bruneau Prospectus and Suggestions," April 1, 1934, typescript, Ross Papers; B. E. Stoutemeyer to C. Ben Ross, May 24, 1935, File 302.12, Box 2/749703, USBRR (qtn.).

63. Wyoming State Planning Board, "Report on the Proposed Diversion of the Waters of Yellowstone Lake," typescript, August, 1937, and "Yellowstone Lake Diversion Report No. 2," typescript, November, 1937, Wyoming State Engineer Papers, WSA; Ise, Park Policy, 143.

64. *Journal of the House of Representatives of the Idaho Legislature, Twenty-fourth Session* (N.p.: n.p., 1937), 182–183, 240, 379.

65. R. W. Faris to James P. Pope, April 4, 1937, IRR.

66. Daniel Tyler, *The Last Water Hole in the West: The Colorado-Big Thompson Project and the Northern Colorado Water Conservancy District* (Niwot: University Press of Colorado, 1992).

67. Cameron, "Irrigation," 42a; R. W. Faris to Barzilla Clark, May 8, 1937, Barzilla W. Clark Papers, ISA; *Ninth Biennial Report...Department of Reclamation*, 27.

68. Charles West to Marvin McIntyre, June 30, 1937, in Nixon, ed., *Roosevelt*, 2:82–83.

69. Franklin D. Roosevelt to James P. Pope, June 29, 1934, in Roosevelt, 1:314–315.

70. Michael P. Malone, *C. Ben Ross and the New Deal in Idaho* (Seattle: University of Washington Press, 1970), 70.

71. Elwood Mead to R. W. Faris, October 13, 1933, IRR; *Tenth Biennial Report of the Department of Reclamation, State of Idaho, 1937–1938* (N.p.: n.p., 1938), 81; U.S. Department of the Interior, Bureau of Reclamation, *Repayment Histories and Payment Schedules–1952* (Washington, DC: Government Printing Office, 1953), 101.

72. National Resources Committee, "Snake River Basin," mimeographed, 1–10, Exhibit B in Cameron, "Irrigation"; "Progress of Investigations of P.W.A. Projects," *The Reclamation Era* (March, 1935), 25:53.

73. B. E. Stoutemeyer to C. Ben Ross, May 25, 1935, Ross Papers.

74. C. Ben Ross to Chase Clark, September 6, 1935, and R. W. Faris to J. P. Pope, April 3, 1935, IRR; Ross to Earl D. Jones, July 6, 1936, Ross Papers.

75. R. W. Faris to Barzilla Clark, n.d., Clark Papers.

76. On proposals for increasing water storage at American Falls, see Haines, *Yellowstone*, 2:346; W. I. Swanton, "Progress of Investigations of Projects," *The Reclamation Era* (September, 1939), 29:233.

77. Progress Planning Report No. 1.5.17–1, mimeographed (Boise: U.S. Department of the Interior, Bureau of Reclamation Region 1, n.d.); B. E. Stoutemeyer to Chief Engineer, October 3, 1935, and Chief Engineer to Stoutemeyer, October 15, 1935, Engineering Correspondence Files, Box 784, USBRR.

78. Ise, *Park Policy*, 433; James Pope to R. W. Faris, March 31, 1938, and Watermaster [Lynn Crandall] to E. V. Sharp, June 13, 1938, IRR.

79. R. W. Faris to James P. Pope, April 4, 1938, IRR.

80. *Salt Lake Tribune*, July 11, 1939, clipping in Albion Charles DeMary Scrapbook, microfilm, ISHS; A. C. DeMary to Paul Nash, June 22, 1939, Pocatello Chamber of Commerce Papers, SC-ISU (qtn.).

81. *Project Data 1981*, 745, 747.

82. "Report to the Congress by the Federal Representative on the Snake River Compact," typescript, n.d., and "Snake River Compact" of October 10, 1949, typescript, Charles A. Robins papers, ISA.

83. Harper, "39 Years of Federal Reclamation in Idaho," 9 (qtn.); John S. Longwell, "The Minidoka Irrigation Project," typescript, n.d., John S. Longwell Papers, AHC; *Twenty-fifth Biennial Report of the Department of Reclamation, State of Idaho, 1966–1968* (N.p.: n.p., 1968), 39.

U.S. Reclamation Bureau. Minidoka Desert in 1905, before irrigation.

Minidoka Desert after irrigation by government. *Library of Congress Prints and Photographs Division, Harris & Ewing Collection, hec 03523 and 03525*

Epilogue

Water, Arid Land, and Visions of Advancement on the Snake River Plain

Idaho's miners and farmers first quarreled about possession of water in territorial times, and such social conflicts persisted even after Idaho legislators enacted laws stating that agricultural uses of water generally took precedence over the desires of other industries to exploit it. Over the next hundred years, access to water remained a pivotal issue when Idahoans attempted to choose ways of exploiting their state's natural resources.

Especially in the southern two-thirds of the state, water has never been far from the minds of the populace. In this region, mining of precious metals declined after 1880, and decades passed before extracting other minerals became economically significant. Forestry, only a slightly larger source of regional income, exhibited limited future economic potential. Manufacturing, aside from small-scale processing of agricultural products, similarly exercised little influence. Within the area stretching over three hundred miles between the Idaho-Wyoming boundary and the Oregon border, water and irrigable land were the principal natural resources that could be developed as a supplement to the region's stock-raising enterprises. With new growth seeming to depend on exploiting land and water, it remained axiomatic for nearly a century that the area's banking, commerce, and small industries prospered if irrigated farming flourished; therefore, putting the waters of the Snake River and its tributaries to work became a cornerstone of economic thinking and social policy in the region. But these conditions added new social and political perplexities when Idahoans tried to decide how best to irrigate the Snake River Plain—a plateau, the largest body of irrigable land in the region, that hugged both sides of the Snake River along its crescent-shaped path across the state.

Early in the twentieth century, a scientist with the United States Department of Agriculture hailed the valleys of the Snake River and its tributaries as an "irrigation belt" containing 5,000,000 acres of potentially arable land. In the forty years prior to his pronouncement, the people of southern Idaho appeared at first glance to have exploited well the Snake River Plain's water and arid land. Settlers established rights under federal laws to about 2,000,000 acres of the land, and they overclaimed by many times the natural flow of water in the rivers. The Boise River, from which irrigation of the surrounding valley began during the 1860s, was overclaimed by millions of acre-feet. Along smaller rivers like the Big and Little Lost, the claims of stockraisers so "overtaxed" streams that "shortage of water" for irrigation became the "rule."[1]

This claiming of water and land obscured the slow progress toward actually making the fullest use of the Snake River Plain's water and land. At the turn of the century, farmers irrigated about 608,000 acres statewide; the bulk of this land was situated at the western and eastern extremities of the plain. In the valleys alongside the Boise and Payette Rivers, irrigated farms occupied approximately 148,000 acres, but five-sixths of the Boise River Valley remained waterless. At the opposite side of the plain, farmers diverted enough water from the Snake to irrigate about 290,000 acres upstream from the American Falls of the river. In the Upper Snake River Valley, farmers irrigated nearly 50,000 acres from small "mountain streams" that flowed onto fringes of the plain. Altogether, this acreage amounted to slightly more than one-half of the area that the United States Geological Survey deemed readily irrigable.[2]

Economic development of arid land and water resources came slowly in part because the state's population remained small, even after it had doubled to 161,772 between 1890 and 1900. In southern Idaho, the most sizable city—Boise—grew to about 6,000 residents in 1900. The region's second most populous center, surrounding Pocatello, grew steadily as well, partly on account of non-agricultural developments; after 1880, the population of the upper reaches of the Snake River Valley finally began to expand significantly as well. Conversely, growth in the southcentral part of the Snake River Plain remained almost at a standstill. The area continued to be grazing domains of groups such as the Sparks and Tinnian cattle outfits, except for scattered settlements like Bruneau, Oakley, and Albion.[3]

Those who came to faster-growing sectors of the southeastern counties and the Goose Creek Valley in southcentral Idaho included many people who had failed to find satisfactory homes and conditions in Utah. To help these people, the Church of Jesus Christ of Latter-day Saints organized colonizing missions to relocate them to farmsteads in Idaho. The same church's authorities similarly encouraged others to better their lots and to "strengthen the cords of the stakes of Zion" on the Snake River Plain.[4] At the western side of the state, other newcomers attempted to build their own utopia at New Plymouth. Meanwhile, seekers of riches rather than human perfection tried arid-land pioneering of the same areas during the 1880s and 1890s. Some of them had fled to the Snake River Plain from the sod-house frontier of Kansas, Nebraska, and the Dakotas to escape droughts and to mitigate the effects of the national depression of 1893–1897 by starting afresh in the Far West.[5]

The new farmers attempted to reclaim arid land cheaply and to limit their out-of-pocket spending through cooperative irrigation schemes. They organized canal companies to manage their irrigation systems; they built their own diversion works at rivers and creeks; by sweat of their brows, they added canals and ditches to the systems; and in return they held shares of canal-company stock and water rights in proportion to their laboring to establish the irrigation systems. Afterward, these canal companies scarcely ever accumulated any capital, the corporations teetered on the edge of bankruptcy, and even small mishaps could drive them out of business. But the cooperative approach to irrigation had allowed farmers to keep most of their money in pocket and drastically reduced irrigation costs regionally. When John Wesley Powell surveyed 4,333 of Idaho's irrigated farms in 1890, he discovered that construction of irrigation systems serving the farms cost on the average $4.74 per acre, and the price of maintaining the systems averaged 80 cents per acre annually.[6]

This cooperative approach to irrigation functioned most satisfactorily in areas where Mormon residents predominated and their cohesive culture and communitarian spirit united the people. However, non-Mormon farmers copied the approach widely. For instance, Payette River Valley farmers organized their own cooperative canal company after private enterprisers had disillusioned them by failing to deliver water as promised.[7]

Dependence on the cooperative approach to irrigation—which continued to spread throughout the Snake River Plain because capitalism repeatedly failed its farm customers—kept farms small, confined them to places close to streams, and left vacant most of the potentially arable land. Everywhere, cooperators lacked resources with which to pay for rock-crib dams or to capitalize more desirable but costly earthen or concrete-arch dams and other diversion technologies that were essential to utilizing copious quantities of water for large-scale irrigation. Instead, at the best, cooperators used waterwheels to lift smaller amounts of water from rivers into canals; more commonly, they positioned their canals to head at sloughs along rivers, then blocked the slough outlets with debris in order to raise stream levels high enough to reach canal intakes. At the worst, a journalist reported that such "shocking [diversion] dams" along the Little Lost River were "built of rubbish, hay, and even stable refuse," and some of them were "patched and caulked with cast-off garments and rags." Spring flooding usually destroyed such structures, and irrigators could replace them with nothing better after the flooding subsided.[8]

By the 1880s private enterprisers claimed anew, despite the dubious track records of their predecessors, that capitalism's practitioners could outperform irrigation's cooperators. Territorial governor Edward A. Stevenson promptly challenged them to reclaim the millions of acres that cooperators had never touched. At the same time, these enterprisers encouraged speculation that, in light of good times nationwide, foreign investors and American railroad magnates and Gilded Age industrialists might capitalize reclamation of the Snake River Plain. The International Immigrant Union and the Union Pacific Railroad also raised expectations for large-scale reclamation in Idaho. The Immigrant Union publicized a scheme to "colonize" 300,000 acres; the Union Pacific also seemingly placed its capital behind arid-land reclamation by sponsoring a Boise and Nampa Canal Company. Visionaries like Arthur Foote appealed to Henry Villard, a Northern Pacific Railroad financier, for underwriting of Foote's blueprints for irrigating the Boise River Valley.[9] Such thinking persisted until the depression of the 1890s dashed hopes for massive infusions of capital into Idaho from either foreign or domestic sources.

After depression had removed all likelihood of access to large pools of outside money, utilization of the Snake River Plain's land and water resources expanded only slowly in spite of the plain's population growth during the 1890s. Entrepreneurs sometimes raised enough money to improve irrigation conditions marginally before overreaching themselves by attempting to capitalize large-scale projects. Some enterprisers, at their best in bettering conditions in the upper reaches of the Snake River Valley, merged cooperative canal companies into fewer but stronger economic units that, in turn, improved irrigation diversion works along the Snake River and placed significantly more water on the land. The least circumspect of them, the Great Western Canal and Irrigation Company (later rechristened Great Western Canal Construction Company), overcharged its customers for land and water; even worse, it sold water that it never delivered to certain lands. Such abuses finally provoked a rebellion by water-users that contributed to the corporation's downfall.[10]

In the Boise River Valley, enterprisers similarly promised more water than they delivered to farmers. For instance, the New York Canal Company carried only trickles of water to a handful of farmers, and even less moisture reached them during drought years. Elsewhere in the same valley, others advanced reclamation of arid land so little that in 1893 Governor William McConnell urged state legislators to spend public funds on building irrigation structures even though erecting the facilities was the obligation of free enterprise—which had incurred such responsibilities by pledging to reclaim arid land on a grand scale.[11]

Whatever their faults, enterprisers defended their new records on grounds that their methods were superior to the cooperators' approach to irrigation. They claimed public policy and social thinking for their side because free enterprise remained the most trustworthy force to spearhead exploitation of the Snake River Plain's land and water. They also argued that faster progress could be expected from free enterprise as soon as the depression of the 1890s ended. But after certain entrepreneurs had misapplied their laissez faire privileges and most of the Snake River Plain remained unirrigated, critics dismissed these arguments and, a decade later, still questioned the integrity of entrepreneurs who engaged in land reclamation. One skeptic hailed the enterprisers' methods as a "siren song" from self-serving boomers; two Idaho state legislators charged that "wildcat" developers had resorted to "fraud" and "crime" in handling irrigation projects throughout the state.[12]

Free enterprise incurred new distrust as a result of its handling of Carey Act reclamation projects during the 1890s. In this approach to land reclamation, which Congress authorized in the Carey Act of August 18, 1894, each of ten arid-land states and territories selected 1,000,000 acres of federal land; then, under state or territorial supervision, free-enterprisers theoretically placed irrigation works on the same land and amassed profits (or at least recovered capital costs) from selling water to settlers. This approach supposedly ensured investment of sufficient money to capitalize large-scale irrigation. However, the results at Idaho's first Carey Act projects (near Marysville and American Falls) reflected adversely on the economic and technological judgment of entrepreneurs; they were no prototypes for large-scale reclamation.[13]

As free enterprise's integrity and skill in reclaiming land became increasingly suspect, state legislators authorized an alternative—a new law that permitted irrigators to create their own irrigation districts. The districts might supplant local canal companies, solicit capital from investors, and pay off such debts by taxing residents of irrigation districts. Irrigators at New Sweden organized a district in 1899; near Caldwell, farmers created the Pioneer Irrigation District in 1901; the Nampa-Meridian Irrigation District emerged in 1904. Altogether, farmers established thirty districts between 1899 and 1916.[14] However, the new districts sated little of the thirst for outside capital for the Snake River Plain. Instead, districts drove unpopular canal owners out of business but, unable to inspire much confidence in the nation's financial marketplaces, raised few dollars. At their worst, the new districts even victimized irrigators by trying to irrigate too much land with limited streams of water. Accusing the Weiser Irrigation District of reckless expansionism in the face of meager water resources, an agrarian stated:

> In the early days we did not experience much trouble in getting ample water...but as the number of acres of [irrigated] land lying to the east of us increased, we can only get water in plenty, when high water season is at hand, and during low water [season], we get practically none.[15]

Depression times ended nationally in 1897, at which point five western railroads—with Great Northern Railroad founder James J. Hill their most eloquent leader—pooled $125,000 to publicize arguments for more irrigated land in the West. For another decade, the railroads persisted in urging Americans to grasp the riches that were said to await holders of irrigated farms.[16] In turn, such advertising renewed old-time dreams of securing money from the Gilded Age's fabled industrialists or the nation's new banker princes with which to reclaim the Snake River Plain. However, no such pools of money became available.

At the turn of the century, southern Idaho stood at a crossroads economically because no other activities had lessened the region's dependence on utilizing its arid land and water. On the other hand, relatively little of this land had been reclaimed during the previous forty years; most of the water in the Snake River and its tributaries escaped irrigators and eventually flowed into the Pacific Ocean. The Idaho State Bureau of Immigration claimed that 9,000,000 acres of land could become irrigable if spring-runoff water from these streams were "properly impounded with [new] dams and reservoirs." Idaho State Engineer Douglas Ross agreed, and he urged the state government to "offer very liberal inducements to capital[izers]" of reclamation lest those acres remain in "desert condition for generations to come." Ross also argued that, given the new prosperity nationally after 1897, the Carey Act approach to reclamation should appeal enough to financiers that they would invest large sums of money in spite of unsatisfactory results on the state's first Carey Act projects.[17] Others were less sanguine about Carey Act methods. They demanded experiments with different approaches to recovering arid land, and their skepticism toward private entrepreneurs spread to the point that little faith remained in the prospect of free enterprise quickly converting the Snake River Plain into farms.

This search for superior ways of reclaiming land continued until, in 1902, a new federal approach to reclamation seemingly guaranteed, at least by first interpretation, to rescue the plain from its aridity. Congress mandated the new approach in the Reclamation Act of June 17, 1902; President Theodore Roosevelt also championed it. Its principles were upheld by conservationist forces deeply entrenched in the executive branch of the federal government that looked to Roosevelt and Gifford Pinchot, head of the U.S. Forest Service, for leadership.

The Reclamation Act of 1902 directed the Department of the Interior to provide irrigation technologies, such as bigger dams across rivers and larger reservoirs of water, because irrigators and their private capitalizers had reclaimed so little land during the previous forty years.[18] The United States Reclamation Service, an arm of Interior, surveyed likely irrigation sites in the West and, in Idaho, narrowed its list of desirable sites to three before finally deciding on two tracts. First, federal planners selected the so-called Minidoka tract; here, on an empty sector in southcentral Idaho, they envisioned irrigation of 120,000 to 150,000 acres. Second, the same planners, who had from the outset eyed the Payette and Boise Rivers' rich water supplies, decided to divert enough water from those streams to irrigate at least 400,000 acres. Part of this land remained untilled in 1902 for want of irrigation water, and the rest was inadequately irrigated by federal standards.[19]

The Reclamation Service's plans for its Payette-Boise and Minidoka projects rekindled old dreams of extracting great wealth from water and arid land, sustaining social advancement regionally, and, in the prophecies of idealists, the erecting of New Canaan on the Snake River Plain. These optimists claimed that signs of such glad tidings had appeared overnight in the form of avalanches of immigrants seeking homesites on the federal reclamation tracts. In 1904, an observer noted, newcomers flocked to the Boise River Valley "by the [railroad] carload"; similarly, settlers crowded onto the Minidoka tract well ahead of significant reclamation of their surroundings for agricultural uses. At Boise, whose population surged from 6,000 residents in 1900 to 18,000 in 1910, so many arrived that the town could not absorb the immigrants this quickly. A local labor union journal reported in 1905: "there are five men for every job in sight there."[20]

Social hardships like Boise's new urban joblessness seemed a tolerable price to pay for socioeconomic advancement, and people residing elsewhere on the Snake River Plain demanded their share of the same progress by duplicating what a Boise River Valley booster described as his area's "glowing future." These forces organized the League of Southern Idaho Commercial Clubs to promote faster population growth across the plain.[21]

The new commercial-club forces helped to lure more immigrants to Idaho, but many newcomers still preferred homesites on the federal tracts to settling on other projects. With newcomers continuously crowding onto the federal projects, lively bartering ensued between them to purchase all privately owned land to which the government might someday supply water. Speculators tended to dominate this trading, and they drove up land prices. In the Boise River Valley, prices for barren land with prospects of water sometime reaching it soared from $50 to $75 per acre. But many other land seekers evaded out-of-pocket spending by practicing squatter sovereignty over "high and dry" public land to which they trusted the government to provide water. In 1905, a journalist said of certain land in the Boise River Valley: "as far as the eye can see the country is dotted with [squatters'] shacks...with here and there a good substantial residence with outbuildings."[22]

Federal reclamation methods accumulated admirers, especially among public officials, merchants, and civic do-gooders, who credited federal methods with effecting more economic progress and social well-being for residents of the Snake River Plain than they had ever enjoyed before. But this acclaim of federal ways waned when it became apparent before 1910 that federal reclaimers could never speedily complete the Payette-Boise and Minidoka projects. The United States Reclamation Service, *Washington Star* publisher William Curtis declared, functioned in a "clam-like" manner.[23] Consequently, owners of land and squatters who had crowded willy-nilly onto these projects endured years of thirst and waiting before receiving any water. Furthermore, the Reclamation Service sidetracked $1,300,000 from the Minidoka project to pursue its work on the Payette-Boise project more quickly. Such shifting of funds delayed until 1909 at the earliest the pumping of water to thousands of Minidoka project acres that were situated at elevations too high to be reached by gravity irrigation methods.[24] The Reclamation Service also changed its plan for carrying Payette River water to farmers, partly because it lacked funds to implement all of its plans at the same time. Next, the agency revised its priorities so that reclaiming land in the Boise River Valley came ahead of Payette River Valley land; at first, too, political pressures on the agency helped induce it to make such changes in its blueprints. Four decades would elapse before Payette River water reached some of the land.[25]

More important, analysts pointed out that, as a result of restricting their work to two projects in the state, federal reclamationists had left most of the Snake River Plain out in the cold (or dry) instead of carving projects from several million acres. Boomers in southeastern Idaho petitioned in vain for federal reconsideration of their proposals for a 200,000- to 300,000-acre Dubois project.[26] Federal reclamation authorities likewise rejected new overtures for placing large projects in the valleys drained by the Big and Little Wood Rivers, in the Big and Little Lost River sectors, and throughout the Twin Falls country. Petitioners claimed that over a million acres beckoned to reclamationists in the Twin Falls area, and one partisan even asserted that federal money spent on building the comparatively small Payette-Boise project could have paid for reclaiming the entire Twin Falls country.[27]

Idaho critics of federalism, nettled by the prospects of federal investment in land reclamation remaining so niggardly, united with other anti-federalists to assail President Roosevelt's conservation policies. They won considerable sympathy, in part by fanning resentment in Idaho over federal schemes for "locking up" irrigable land in a Shoshone Falls National Park. They likewise upheld Governor Frank Gooding in his charges that "conditions in Idaho" stayed "much disturbed" because Gifford Pinchot and his Forest Service agents abused their powers in carving out national forest reserves. In one instance, a critic of the Forest Service complained, the agency almost succeeded in placing virtually unforested parts of four Idaho counties within a forest reserve.[28]

At the same time, these critics of federal conservation policies accused Pinchot and Frederick Newell, the like-minded chief of the United States Reclamation Service, of padlocking the West's natural resources in order to create empires for themselves to administer. As a measure of Pinchot's and Newell's success, an Idaho State Engineer charged, western states no longer exercised "right of control of their own natural resources." Another Idaho official protested against Newell's purported monopoly by demanding that private enterprise at least receive a "fair chance" to reclaim land that Newell's agency could not touch for "a good while."[29] To such criticism of the Reclamation Service, the Roosevelt administration retorted that the agency did jobs at which private enterprise had failed. In 1909, Secretary of the Interior Richard Ballinger reiterated these claims and stated that federal reclamationists only stepped in after free enterprise had completed "what it can."[30]

Scoffing at such defenses of federal land-conservation policies, many Snake River Plain boosters underscored their disappointment with the government's land reclamation efforts by noting again that the Reclamation Service's course appeared to preclude it from ever effecting full utilization of the plain's land and water. Accordingly, these boosters looked once more for non-federal ways of realizing their dreams; and, fortuitously for them, private enterprise showed more convincingly than ever how large-scale reclamation under the Carey Act was practicable despite entrepreneurial shortcomings of the past. In an especially persuasive demonstration of free enterprise's latest arguments for its side, Pennsylvania entrepreneurs Peter Kimberly and Frank Buhl hit the jackpot after 1902 in the heart of the Twin Falls country. First, they raised over $3,000,000 with which to capitalize their 244,000-acre Twin Falls South Side project; next, in less than a decade, Buhl completed this work to the satisfaction of public authorities who supervised Carey Act projects within the state. In the end, nearly 200,000 South Side acres were actually reclaimed.[31]

Admiration of Buhl's quick results mounted and, in turn, helped to inspire greater public confidence in Carey Act reclamation methods. A state official claimed that by 1908 "almost as much" land reclamation had been "accomplished" on the South Side project as had been achieved in the Boise River Valley over the previous thirty-five years.[32] A *Harper's Weekly* writer rated the South Side project among the "miracles" of the times. In 1907, an observer stated that buyers of South Side land had already "made good money," and a local jurist described the project as "the certain prophecy of a prosperous civilization."[33]

During the same years that "miracles" reportedly unfolded on the Twin Falls South Side tract, Progressive Age reformers, publishers of newspapers like the *Chicago Tribune*, publicists for western railroads, and certain of the nation's literati strengthened the case for adopting Carey Act reclamation methods. Irrigation, they claimed, resulted in accumulating riches from changing western "lava dust" to "gardens"; such arguments legitimized nationally the financing of reclamation projects in essentially the same ways by which Buhl and Kimberly had secured underwriting for their project. In this approach to capitalizing reclamation, irrigation companies won authority to create projects, supposedly under the watchful eyes of state authorities who administered the Carey Act within their respective jurisdictions. The companies then lured investors to financial marketplaces that specialized in marketing their issues of stocks and bonds; once investors were tempted, brokerages such as Trowbridge and Niver of Chicago unloaded the irrigation companies' securities on them. The publisher of *Financial Publications*, a Boston-based newsletter, rated this paper "a lot of improperly secured bond issues," but other commercial journals cautiously endorsed the irrigation securities. As in all legitimate marketplaces, Chicago-based *Bonds and Mortgages* stated, buyers of irrigation-company securities needed simply to distinguish between "good" and "bad" paper in order to avoid purchasing the "counterfeit" instead of the "genuine."[34]

These marketplaces attracted principally investors of modest sums of money, and many Snake River Plain boosters doubted that such small fry could subscribe enough money to underwrite filling the plain with irrigated farms. They believed that appealing somehow to industrial princes and the nation's wizards of finance capitalism would be a more effective way of capitalizing large-scale reclamation of land. However, these doubters chose not to look gift horses in their mouths: to do so would defy common sense, after Kimberly and Buhl had collected from small investors enough dollars to sustain their Twin Falls South Side "miracle." Furthermore, Idaho's Board of Land Commissioners (commonly called the Land Board) alleviated some misgivings in the state about depending on such investors for capital. The board, a public agency that supervised Carey Act projects in Idaho, encouraged investors of modest means to take their chances on Carey Act projects instead of spending their funds on different types of irrigation ventures.[35] Spokesmen for Carey Act interests assured investors that listening to the Idaho Land Board was sensible because only "reasonable business people" sat on it.[36]

In response to such importuning, investors from middle-class groups like physicians, engineers, and merchants bought enough irrigation-company securities that architects of Idaho's Carey Act projects learned to depend on them for increasingly larger subscriptions of capital. By 1910, the entrepreneurs were emboldened enough to count on this class of investors for most of their capital needs of $68,479,015. A year later, they revised their plans upward; altogether, they now expected to spend $75,667,540 on forty-two projects for which they had received the Land Board's authorization.[37] All in all, the state's Carey Act Land Commissioner estimated that opening up such projects brought $100,000,000 of outside wealth into Idaho between 1898 and 1913.[38]

Given what seemed credible promises from private entrepreneurs, trusting the Snake River Plain's fate to free enterprise gained widespread social acceptance during the decade ending in 1912. In these times, the plain's boosters argued that previous shortcomings of free enterprise at places like the Marysville project deserved forgiveness and that, at the hands of newer enterprisers, achievement of what had always been deemed the region's economic destiny and social well-being was in sight. They exulted for good reason. Entrepreneurs eventually petitioned the Land Board for authority to undertake sixty-five new projects. The more exciting of these proposals envisioned irrigated farms throughout areas that federal reclamationists had bypassed but were believed suited to large-scale reclamation. The sites for such new projects included over 160,000 acres in the Wood River valleys, somewhat smaller irrigation developments in the Big and Little Lost River sectors, and about 1,000,000 acres surrounding the Twin Falls South Side tract. Governor Gooding insisted that Idaho accommodate all projects which the Land Board judged worthy; accordingly, the state acquired from the federal government another 2,000,000 acres to be developed under Carey Act processes.[39]

Perceptions of richer times just around the corner generated new interest, immigration of more people, and expansion of irrigated farming even in sectors of the Snake River Plain where formation of Carey Act projects lagged. Particularly in the upper reaches of the Snake River Valley, an area in which creation of Carey Act projects came to a standstill except at Mud Lake and Marysville, the population grew rapidly between 1900 and 1910. At Blackfoot, for instance, it increased from 900 to 2,200.[40] Elsewhere, entrepreneurs mapped out grander Carey Act projects now that prospective settlers came in greater numbers and plenty of public land was at the state's disposal to accommodate them. At one point, the most visionary of these schemes called for a 527,000-acre Big Bruneau project in Twin Falls and Owyhee counties, only for James and William Kuhn to draft plans for topping their competitors' Big Bruneau undertaking. By one estimate, Carey Act reclamation activities alone lured 50,000 people to Idaho before 1913.[41]

Noting such advances, analysts declared conquest of the Snake River Plain for agricultural ends nearly accomplished. But in 1911 this boom in recovering arid land suddenly faltered, and conditions worsened so fast that the boom collapsed within another year. First, in what one publicist called a "disgraceful episode," the Big Lost River Irrigation Company went bankrupt, leaving buyers of its bonds with little prospect of recovering their investments of $1,355,000. Bankruptcies next engulfed more corporations that were engaged in reclaiming western land, and brokerages like Trowbridge and Niver of Chicago, which once acted only as middlemen between irrigation companies and investors, fell by the wayside as a result of trying belatedly to finance irrigation ventures in Colorado, Idaho, and Oregon. Finally, in 1913, financial disasters overwhelmed the banking and industrial empires of William and James Kuhn. The Kuhn brothers had most recently attempted to superintend reclamation of about 500,000 acres in southcentral Idaho.[42]

It was no wonder at all that confidence in western irrigation securities declined to the point that old marketplaces for such commercial paper rejected it; distrust of proposals for western irrigation ventures pervaded other investment-capital markets, and public figures shared in this new discountenancing of land reclamation. In remarks before the Lotus Club of New York City, Ohio governor Judson Harmon steered investors away from the West's irrigation enterprises. He said that, in the northwestern states, it was "untold wealth of [water] power and treasure of mine, quarry, and forest" which "await[ed] the magic touch of capital."[43]

This investor abandonment of irrigation-company stocks and bonds tended to discredit Carey Act approaches to large-scale irrigation in Idaho. At the least, the consequences of investors' disenchantment represented formidable setbacks; up to 1912, purchasers of irrigation-company bonds subscribed most of $18,000,000 that had paid for irrigation works serving the earlier Carey Act tracts, but the same class of buyers subsequently withheld investment of $58,000,000 in needed capital from projects that had received approval from the Land Board. Probably in greater measure, the post-1912 backlash in Idaho against Carey Act methods reflected social frustration that, after rosy forecasts, conquest of the Snake River Plain by irrigation was not in sight. Out-of-state financial support had vanished as well. Promoters of the Big Bruneau project led the pack in applying to Wall Street bankers for enough capital to complete the more promising of the unfinished Carey Act projects. The bankers rebuffed them.

In short, as of 1920, state authorities considered only about 860,000 of the state's 3,000,000-acre Carey Act land grant to be permanently irrigable. According to the Idaho Commissioner of Reclamation's calculations, this land received water through irrigation works built at a cost of $25,413,000 (a part of which money the bondholders of old Carey Act irrigation companies and Idaho irrigators had necessarily dug from their own pocketbooks after 1913). Beginning in 1921 and 1922, entrepreneurs commenced no new Carey Act projects in the state for the first time since the 1890s.[44]

Except for the federal government's Payette-Boise and Minidoka projects, the surviving irrigation tracts fitted into a jumble of irrigation districts, projects operated by old-time canal companies (now largely farmer-owned corporations), and Carey Act projects (in which a farmer-controlled canal company eventually managed the irrigation works at each tract). Authorities in these jurisdictions attempted to hold as much as possible of the ground that had been gained in better times.[45]

Many Snake River Plain boosters regarded such status quo conditions as unacceptable. They blamed free enterprise for the reclamation boom collapsing; they resisted continued reliance on Carey Act methods but vowed to recover for irrigation its old expansionism.[46] However, they disagreed on how as much as 9,000,000 acres could be reclaimed, so public authorities tried out several possible solutions at the boosters' behest. In what seemed the most promising answer to the malaise, state legislators tinkered with irrigation-district laws in the hope that changing the statutes would attract capital from investors who might loan money to irrigation districts but withhold it from irrigation companies. By 1921, legislators believed their newest handiwork made irrigation-district bonds irresistible to investors. The lawmakers were mistaken.[47] Meanwhile, Governors John Haines (1913–1915) and Moses Alexander (1915–1919) demanded federal assistance in resuscitating reclamation activity at several of the most trouble-plagued Carey Act projects. Alexander finally persuaded federal authorities to take charge at the King Hill tracts.[48]

When these strategies never significantly changed conditions, with the consequence that no more dams and reservoirs or newly irrigated land materialized, skeptics challenged notions that public policy, economic expectations, and social priorities must unwaveringly uphold filling the Snake River Plain with irrigated farms. They urged that the region subsist on the irrigation technologies acquired before the reclamation bubble burst. The group also believed that, with better management of the plain's river systems, such technologies sufficed for irrigating additional acres without depriving farmers of adequate quantities of water. This plan likewise envisioned large social dividends when better management of streams eliminated conflicts between the users. Experiments along this line culminated in 1923 in formation of Irrigation District Number 36, through which delivery of Snake River water to areas stretching from Ashton to Twin Falls was managed. Meanwhile, agricultural scientists and certain public officials advocated moisture-conservation strategies in order to force water at hand in reservoirs to serve more acres; raising capital to expand the existing irrigation systems would thus be less essential. Such capital-raising efforts seemed foolhardy after Idaho's Carey Act reclamation experiment "blew up with a bang," and reclamation activity in the state suffered permanently from a "black eye."[49]

Storms of disapproval swirled around such schemes to reduce expectations. Farmers understandably opposed dividing limited quantities of water among larger numbers of irrigators. No less uncompromising, civic boosters and commercial forces reiterated old demands for enough dams and reservoirs so that the last drops of water would turn more tracts of land green. To reach this end, a publicist for the Idaho Reclamation Association stated, Idahoans were prepared to accept not "theory" but "cold cash" from private enterprise to accomplish a "wedding of the land and water" of the Snake River Plain; what these forces really wanted, United States Senator John Nugent observed, was federal construction of sufficient dams and reservoirs to sustain the marriage. Governor David Davis (1919–1923), who also believed federal intercession more likely to save the day, launched the Western States Reclamation Association to lobby Congress for bigger appropriations for reclamation purposes.[50]

Notwithstanding Davis' efforts and the dogmatism of groups calling for reclamation of huge bodies of land, evidence mounted during the 1920s against such ideas. According to a United States Reclamation Service engineer, not enough water could be impounded from the Snake River and its tributaries to sustain irrigation of such magnitude; the more pessimistic administrators in the state Department of Reclamation calculated that, by impounding all water upstream from Milner Dam (the mainstay of the Twin Falls South Side project), another 1,125,000 acres might be irrigated at reasonable cost to farmers. The visionaries' case for water aplenty on the Snake River Plain was belied when, to save certain Carey Act projects from collapsing, tens of thousands of acres were deleted from the tracts on account of water insufficiency. For instance, state officials and courts scaled back the Twin Falls North Side project from 261,945 to 185,000 acres.[51]

Ancient nostrums for achieving good times on the Snake River Plain became even less popular when dry seasons during the 1920s highlighted the plain's relatively scarce water resources as compared to its plentitude of arid land, and the nostrums finally fell by the wayside after severe droughts successively in 1919, 1924, and 1933–1934 seemed to dash all hopes for much additional greening of the plain. The droughts showed that for all irrigated sectors—even at the Twin Falls South Side and Minidoka projects, which reputedly possessed ample water resources before droughts showed otherwise—prudence dictated that irrigators accumulate extra water during wet years to tide them over dry years. Accordingly, farmers on these tracts demanded for themselves all unappropriated water in the Snake River and its tributaries. State and federal authorities supported their wishes; the Idaho Commissioner of Reclamation explained that irrigation of "new land" was now discouraged, and public policies instead aimed at "bringing success to settlers on existing projects." Moreover, New Dealers came to town during the 1930s with lots of money to spend, some of it on land reclamation work in Idaho; but, Congressman Compton White complained, they consistently denied such funds to "new projects or any new land."[52]

A remnant of the old-time forces for banishing aridity from the Snake River Plain, acknowledging the plain's limited water resources, refused to give up their cause. These visionaries proposed diverting water to the region from the Salmon River, from Montana's Madison River, and from lakes and streams within Yellowstone National Park. Not surprisingly, their plans failed—except that for tapping a small pool of water at Grassy Lake within Yellowstone National Park.[53]

For most irrigators the decades from 1920 to 1960 were times to find less dramatic, fiscally more sensible, and environmentally more acceptable ways of coping with the Snake River Plain's inadequate supply of water. They chose eventually to depend on the plain's own water resources. Implicitly, this approach signaled both the final retreat in Idaho from grander economic and social expectations for the plain and an effort to salvage as much as possible of irrigated tracts—no matter their origins—that were launched in days when unbridled progress was taken for granted. A first long stride in this direction, a new dam and reservoir at American Falls, augmented water supplies and finally bequeathed greater security to several irrigated tracts in southcentral Idaho after 1927. The American Falls facility also permitted diversion of Snake River water, via a seventy-mile canal heading at Milner Dam, to the water-deficient Wood River valleys (although the sector depending solely on Magic Reservoir continued for the rest of the century to need more water than was available to it during dry years).[54]

Next, new dams and reservoirs at Island Park and Palisades significantly increased the water supply for about 670,000 acres and relieved distress that historically had troubled the region upstream from American Falls. Before the new structures served the area, in many years irrigators could not store enough of the Snake River's spring runoff water to sustain them through the summer. Even in times of no real drought, quirks like unseasonably warm weather during spring and early summer rains, as happened in 1936, had resulted in "waste" of runoff water that farmers needed later in the season but could not reserve before it escaped downstream.[55]

Also in these times of accepting lesser results from land reclamation efforts, federal builders placed new dams and reservoirs to serve the Boise and Payette River valleys at Black Canyon (1924), Deadwood (1931), Cascade (1948), Anderson Ranch (1950), and Lucky Peak (1952)—an Army Corps of Engineers flood-control, rather than a Bureau of Reclamation, project. The structures curbed spring flooding of the valleys, impounded extra water that momentarily sated the appetites of Boise River Valley farmers for moisture, and brought the federal government's Payette-Boise project to completion. At this point, the Payette-Boise project contained 390,000 acres. The new irrigation works also provided sufficient water to rescue Payette River Valley farmers who had earlier trusted the Canyon Canal Company to deliver water to them under the Carey Act. Electricity generated at Black Canyon Dam kept the Gem Irrigation District's pumps running until the district's problems were more permanently addressed by including it within the Owyhee project, a federal reclamation undertaking in Oregon and Idaho.[56]

Although not advancement of a magnitude that visionaries once expected, securing the new dams and reservoirs along the Snake, Boise, and Payette Rivers extended irrigated farming on the Snake River Plain, ensured the hegemony of agriculture regionally, and contributed to the advance of farming past mining and lumbering as the state's greatest economic asset.[57] Construction depended heavily on federal largesse, and the United States Bureau of Reclamation provided such structures for irrigators on terms affordable for farmers. The federal Reclamation Act of 1939 offered better financial and tenurial conditions to these irrigators; the statute also lightened the burdens on them by apportioning part of the construction costs of dams and reservoirs to buyers of electricity generated at the dams and to beneficiaries of flood control on the same rivers. In addition, during the decades from 1920 to 1970, the federal reclamation bureau provided storage reservoirs for several small projects that depended for water on Snake River tributaries. Finally, the bureau placed a major dam across the Teton River—only to have the structure collapse on June 5, 1976.[58]

During the post-World War II decades, the irrigated acreage of Idaho rose to about 4,000,000, a portion of which became irrigable without adding more dams and reservoirs to irrigation infrastructures. Instead, because of technological improvements, it became economically feasible to pump water onto this land from surface streams and underground sources. For instance, pumping led to the farming of another 76,796 acres situated in the area adjacent to the original Minidoka project. Elsewhere, too, usage of pumping technologies continued to grow. In 1989, the Idaho Power Company reported, 47,718 acres of cropland were newly watered by use of electrically driven machinery.[59]

In these times when progress again marked irrigation developments on the Snake River Plain, the grandest of old visions for greening up the plain seemed in retrospect to be misguided. The Big Bruneau and Dubois project schemes became only memories; and, despite sporadic discussions of somehow irrigating the Mountain Home plateau, it remained as arid as ever. Schemes for diverting water from the Salmon River and Yellowstone National Park were now classed alongside the other pipe dreams of earlier generations about extracting riches from the Snake River Plain. Then, in the 1970s, the Idaho Department of Water Resources pessimistically assessed the prospects for more lessening of aridity on the plain. According to the department, the plain contained another 7,400,000 acres of land that could benefit from irrigation were water at hand for it, but slightly less than a million of those acres might in fact become irrigated cropland during the next fifty years.[60]

Lowering of socioeconomic expectations from the Snake River Plain became inevitable when even diehards realized that, instead of placing public priorities on launching new reclamation projects after World War II ended, a struggle lay ahead to keep green the land that had been transformed from its arid condition during the previous half-century. Eventually, as a federal Bureau of Reclamation official warned, irrigation systems aged to the point that routine maintenance processes no longer sufficed, and "rehabilitation and [probably] replacement of existing irrigation systems" were required. Hence, during the four decades following the conclusion of World War II, sustaining the plain's irrigation infrastructures ranged from rebuilding the American Falls Dam and arranging to refurbish Milner Dam to replacing smaller irrigation works on projects like the King Hill tract. Furthermore, a federal official charged in 1990, serious disrepair had affected other dams because, in Idaho, public decision-making processes typically responded not to "near misses" but to "the squeaky wheel or the body bags."[61]

New Deal programs, the United States' national security needs during World War II, and the country's cold war with the Soviet Union after 1945 induced changes across the Snake River Plain that undermined the Potato Alley perspectives which had previously underlain thinking about the plain's fate. New Deal programs encouraged industrialization in the West, and America's hot and cold wars resulted in the establishment of the Idaho National Engineering Laboratory and military installations. Over the long haul, these institutions supplied growing numbers of people with long-term ways other than agriculture to make a living. The quickening of commerce in southern Idaho after World War II and industrialization at the hands of the FMC Corporation and others made available yet more non-agricultural means of earning livelihoods. In turn, this new commercialism and industrialism introduced different concepts of socioeconomic progress based on using the Snake River Plain's land and water in ways other than for irrigated farming.[62]

In the end, these newest developments generated little agreement on ways of utilizing the Snake River Plain's arid land and water. Out-of-staters maneuvered to take away the water. Manufacturers of electricity and new-age industrialists demanded the same water for their commercial purposes; but nature's preservationists, environmental activists, and urban sportspeople counter-proposed their own visions of how better to use land and water in achieving the plain's socioeconomic destiny. Those forces preferred placing the land and water beyond the grasp of both industry and agriculture.[63]

The process of recharting the region's social ideologies for progress, prosperity, and environmental sensitivity was under way at the end of the twentieth century. In this process of redesigning blueprints for best using the plain's most conspicuous natural resources, non-agrarian forces threatened to shoulder aside the irrigators from their historic advantage of having first-in-line access to arid land and water. In short, even after a century of debate in Idaho, choosing how best to utilize the Snake River Plain's land and water remained an issue to be decided.

This article originally appeared in *Idaho Yesterdays* 35, no. 1 (Spring 1991). Reprinted with permission.

Notes

1. James Stephenson Jr., *Irrigation in Idaho*, U.S. Department of Agriculture Bulletin 216 (Washington, DC: Government Printing Office, 1909), 23 (1st qtn.); Annie Laurie Bird, *Boise, The Peace Valley* (Caldwell, ID: Caxton Printers Ltd., 1934), 200–203, 272–78; Richard Calvin Montgomery, "Canyon County: The Economic Geography of a Southwestern Idaho Irrigated Area" (master's thesis, University of Nebraska, 1951), 37–38; *Biennial Report of the State Engineer to the Governor of Idaho for the Years 1899–1900* (Boise: Capital Printing Office, n.d.), 26–28.

2. Oscar Osburn Winther, *The Great Northwest: A History* (New York: Alfred A. Knopf, 1947), 273–74; Merrill D. Beal and Merle W. Wells, *History of Idaho*, 3 vols. (New York: Lewis Historical Publishing Company, 1959), 2:184; *Biennial Report of the State Engineer...1899–1900*, 14; E. B. Darlington, "Irrigation Survey of the Upper Snake River Valley, Idaho," *Engineering News* (June 1, 1905), 53:576 (qtn.).

3. Dorothy O. Johansen and Charles M. Gates, *Empire of the Columbia: A History of the Pacific Northwest*, 2nd ed. (New York: Harper and Row, 1967), 607; Boise Centennial Committee, "Boise, City of Trees: A Centennial History," Idaho Historical Series Number 12 (Boise: Idaho State Historical Society, 1963); Robert L. Wrigley Jr., "The Early History of Pocatello, Idaho," *Pacific Northwest Quarterly* (October 1943), 34:360–361; *Biennial Report of the State Engineer...1899–1900*, 52–54.

4. Leonard J. Arrington, *Great Basin Kingdom: An Economic History of the Latter-day Saints* (Cambridge, MA: Harvard University Press, 1958), 30, 354–55; Leslie L. Sudweeks, "Early Agricultural Settlements in South Idaho," *Pacific Northwest Quarterly* 28 (April 1937):149–150; D. W. Meinig, "The Mormon Culture Region: Strategies and Patterns in the Geog-

raphy of the American West, 1847–1964," *Annals of the Association of American Geographers* 55 (June 1965):207; Merrill D. Beal, *Intermountain Railroads, Standard and Narrow Gauge* (Caldwell, ID: Caxton Printers, 1962), 125 (qtn).

5. *New Plymouth Colony Co., (Ltd.)* (N.p: n.p., n.d. [1896]), copy in New Plymouth Townsite Company and C. M. McBride Collection, Idaho State Historical Society, Boise (hereafter ISHS); William E. Smythe, *The Conquest of Arid America* (1899, 1905; repr., Seattle: University of Washington Press, 1969), 131–132, 192–193; "[Reminiscences about] Really Hard Times," *Shoshone Journal*, July 24, 1908, 5. The New Plymouth Colony originated in discussions at the National Irrigation Congress in Chicago in 1894.

6. J. W. Powell, *Thirteenth Annual Report of the United States Geological Survey to the Secretary of the Interior, 1892–93* (Washington, DC: Government Printing Office, 1893), 124. For accounts of such cooperative irrigation ventures, see: *Biennial Report of the State Engineer...1899–1900*, 18; Walden Fawcett, "Irrigation in Idaho," *Scientific American* 83 (September 8, 1900):149; Kate B. Carter, *Pioneer Irrigation, Upper Snake River Valley* (N.p.: Daughters of Utah Pioneers, 1955), 12–14, 30, 67–86.

7. Leonard J. Arrington and Dean May, "'A Different Mode of Life': Irrigation and Society in Nineteenth Century Utah," in *Agriculture in the Development of the Far West*, James H. Shideler, ed. (Washington, DC: Agricultural History Society, 1975), 3–20; Claire Goldsmith, ed., *In the Shadow of the Squaw* (Caldwell, ID: Caxton Printers, 1953), 24–26.

8. Mary Gunnell Lewis, "History of Irrigation Development in Idaho" (master's thesis, University of Idaho, 1924), 31; Carter, *Pioneer Irrigation*, 10–11; William Darrell Gertsch, "The Upper Snake River Project: A Historical Study of Reclamation and Regional Development, 1890–1930" (PhD diss., University of Washington, 1974), 34; *Shoshone Journal*, August 7, 1908, 1 (qtns.).

9. Gilbert C. Fite, *The Farmers' Frontier, 1865–1900* (New York: Holt, Rinehart and Winston, 1966), 189; Davis Bitton, "Blackfoot: The Making of a Community," *Idaho Yesterdays* 19 (Spring 1975), 1:14; Robert G. Athearn, *Union Pacific Country* (Chicago: Rand McNally and Company, 1971), 386; "Memoir of Arthur DeWint Foote," *Transactions of the American Society of Civil Engineers* (1934), 99:1499; Rodman W. Paul, ed., *A Victorian Gentlewoman in the Far West: The Reminiscences of Mary Hallock Foote* (San Marino, CA: The Huntington Library, 1972), 30–31, 302–306, 327–28.

10. Carter, *Pioneer Irrigation*, 108–113; [Merle Wells], *A Short History of Idaho* (Boise: Idaho State Historical Society, 1974), 79; Henry Winter to J. H. Hawley, n.d., James H. Hawley Papers, Governors' Files, Idaho State Archives, Boise (hereafter ISA). Within a decade after 1900, irrigators in the upper Snake River Valley utilized virtually all of the natural flow of the Snake River except in periods of unusually high runoff of water: *Ninth Biennial Report of the State Engineer to the Governor of Idaho, 1911–1912* [N.p.: n.p., n.d.], 337–38.

11. Paul L. Murphy, "Early Irrigation in the Boise Valley," *Pacific Northwest Quarterly* 44 (October 1953):181; [Wells], *Short History of Idaho*, 78; F. Ross Peterson, *Idaho: A Bicentennial History* (New York: W.W. Norton & Company, 1976), 125–27; Neil H. Carlton, "A History of the Development of the Boise Irrigation Project" (master's thesis, Brigham Young University, 1969), 15–16.

12. Gertsch, "Upper Snake River Project," 114–15; *Twin Falls News*, January 26, 1906, 4; *Caldwell Tribune*, July 27, 1907, 4 (1st qtn.); *Rupert Pioneer-Record*, March 4, 1909, 7 (all other qtns.).

13. For a history of the Marysville project, see Patricia Lyn Scott, "Idaho and the Carey Act, 1894–1930: Reclamation by the States" (master's thesis, University of Utah, 1983), 101–32;

and for brief references to these projects, see Mikel H. Williams, *The History of Development and Current Status of the Carey Act in Idaho* (Boise: Idaho Department of Reclamation, 1970), 16–18, 52–53; Hugh T. Lovin, "The Carey Act in Idaho, 1895–1925: An Experiment in Free Enterprise Reclamation," *Pacific Northwest Quarterly* (October 1987), 78:124 [Chapter 4 in this book].

14. Lewis, "Irrigation Development," 56–57; State Engineer to R. N. Hill, March 15, 1917, Idaho Reclamation Records, ISA (hereafter IRR).

15. F. Zlabek to James Stephenson Jr., March 20, 1906, IRR.

16. Paul W. Gates and Robert W. Swenson, *History of Public Land Law Development* (Washington, DC: Public Land Law Review Commission, 1968), 650; Randall R. Howard, "Following the Colonists: An Account of the Great Semi-Annual Movement of Homeseekers," *Pacific Monthly* 23 (May 1910):531–32. See also *Caldwell Tribune*, December 22, 1900, 1; *Twin Falls News*, December 23, 1904, 5; *Rupert Pioneer-Record*, September 2, 1909, 1.

17. *The State of Idaho: Official Report of the Bureau of Immigration, Labor and Statistics* (N.p.: n.p., n.d. [1906]), 40 (1st qtn.); D. W. Ross to State Board of Land Commissioners, October 12, 1900, IRR (2nd and 3rd qtns.).

18. For the triumph of ideas that held federalization of reclamation appeared to be the only way of effecting widespread reclamation in the West, see Roy M. Robbins, *Our Landed Heritage: The Public Domain 1776–1970* (2nd rev. ed., Lincoln: University of Nebraska Press, 1976), 325–333; Donald Worster, *Rivers of Empire: Water, Aridity, and the Growth of the American West* (New York: Pantheon Books, 1985), 150–69.

19. D. W. Ross, "Summary of Investigations in Idaho," January 26, 1904 (typescript), File 123, Bureau of Reclamation Records, Record Group 115 (hereafter cited BRR), Federal Archives and Records Center, Denver; *Caldwell Tribune*, December 19, 1903, 1; "History of the Minidoka Project, Idaho to 1912," 1:4–8 (typescript), Project Histories and Reports of Reclamation Bureau Projects: Minidoka Project, Idaho, 1910–1919, BRR, Microcopy 96, Roll 96, National Archives; "History, 1902–11," 15 (typescript), Project Histories and Reports of Reclamation Bureau Projects: Boise Project, Idaho, 1906–1912, BRR, Roll 8 (hereafter "Project Histories, Minidoka," and "Project Histories, Boise").

20. Ernest G. Eagleson to Frank S. Davis, March 11, 1904, Ernest G. Eagleson Papers, ISHS (1st qtn.); "Boise, City of Trees," *Boise Unionist*, n.d., as cited in *Emmett Index*, April 6, 1905, 5 (2nd qtn.).

21. Ernest G. Eagleson to Frank C. Manley, December 7, 1904, Eagleson Papers, (qtn.); *Twin Falls News*, March 23, 1906, 1; *Shoshone Journal*, December 20, 1907, 1; *Rupert Pioneer-Record*, October 15, 1908, 1 (supplement).

22. *Fund for Reclamation of Arid Lands: Message from the President of the United States Transmitting a Report of the Board of Army Engineers in Relation to the Reclamation Fund*, House Document 1262, 61st Cong., 3d sess. (January 6, 1911), 55; "Idaho: Boise Project and Related Features" ([Washington, DC]: Department of the Interior, United States Reclamation Service, July 1, 1920), 21 (mimeographed), copy in Burton L. French Papers, Miami University Library, Oxford, Ohio; *Emmett Index*, March 16, 1905, 5 (qtns.).

23. Curtis' comments as cited in *Twin Falls News*, December 22, 1905, 7. As of 1910, federal reclaimers supplied water to 38 percent of Minidoka project land; the proportion for the Payette-Boise project was 10 percent. The proportions of such irrigated land almost doubled by 1913 on both projects: Ray P. Teele, *The Economics of Land Reclamation in the United States* (Chicago: A. W. Shaw Company, 1927), 188; R. P. Teele, *Irrigation in the United States:*

A Discussion of Its Legal, Economic and Financial Aspects (New York: D. Appleton and Company, 1915), 72.

24. "Minidoka Project to 1912," 1:27; "Notes for Senator [Fred] Dubois," June 15, 1905 (typescript), File 123, BRR; H. H. Caldwell and Merle Wells, *Economic and Ecological History Support Study for a Case Study of Federal Expenditures on a Water and Related Land Resource Project: Boise Project, Idaho and Oregon* (Moscow: Idaho Water Resources Research Institute, 1974), 45.

25. D. W. Ross, "Work of the Reclamation Service in Idaho," *Pacific Monthly* 16 (September 1906):316; F. H. Newell to Secretary of the Interior, March 26, 1910, Project Histories, Boise; Walter R. Cupp to John C. Page, December 29, 1936, File B420.1, BRR; *Project Data 1981*, Water and Power Resources Service, U.S. Department of the Interior (Washington, DC: Government Printing Office, 1981), 47. For an account of numerous conflicts of Payette-Boise and Minidoka project settlers with the U.S. Reclamation Service, see Charles Coate, "Federal-Local Relationships on the Boise and Minidoka Projects, 1904–1926," *Idaho Yesterdays* 25 (Summer 1981), 2:2–9.

26. For early federal consideration of the Dubois project, see D. W. Ross, "Summary of Investigations in Idaho," January 26, 1904 (typescript), File 123, BRR; and for a sketch of the irrigation scheme for the Dubois project, see W. G. Hoyt, *Water Utilization in the Snake River Basin*, Water Supply Paper 657, Geological Survey, U.S. Department of the Interior (Washington, DC: Government Printing Office, 1935), 175.

27. Paul S. A. Bickel, "Bruneau," n.d. (typescript), David W. Davis Papers, Governors' Files, ISA.

28. D. W. Ross to State Board of Land Commissioners, October 12, 1900, IRR (1st qtn.); F. R. Gooding to Gifford Pinchot, July 5, 1906, Frank R. Gooding Papers, Governors' Files, ISA (other qtns.); Edgar Wilson to W. D. Heyburn, March 21, 1903, Weldon B. Heyburn Papers, University of Idaho Library, Moscow. Because of outcries in the western states, the U.S. Forest Service deleted thousands of acres of potential agricultural land from its forest reserves in 1907: Robbins, *Our Landed Heritage*, 349–351.

29. A. E. Robinson to J. Arthur Eddy, August 18, 1911 (1st qtn.), and Idaho State Engineer to John H. Wade, November 15, 1906 (other qtns.), IRR. For western opposition to other federal controls restricting access to land resources, see John Opie, *The Law of the Land: Two Hundred Years of American Farmland Policy* (Lincoln: University of Nebraska Press, 1987), 149–150.

30. Ballinger's comment as cited in James Penick Jr., *Progressive Politics and Conservation: The Ballinger-Pinchot Affair* (Chicago: University of Chicago Press, 1968), 61.

31. Williams, *Carey Act in Idaho*, 69, 71; Beal and Wells, *History of Idaho*, 2:139–140. An adequate history of the Twin Falls South Side project remains to be written, but the following contain relevant materials: Byron Hunter and Samuel B. Nuckols, *An Economic Study of Irrigated Farming in Twin Falls County, Idaho*, United States Department of Agriculture Bulletin No. 1421 (Washington, DC: Government Printing Office, 1926); Twin Falls County Centennial Committee, *A Folk History of Twin Falls County* (Twin Falls: Standard Printing Company, 1962); Lloyd E. Byrne, *Buhl As It Was* (Boise: Syms-York Company, 1976).

32. "Operations under the Carey Act in Idaho and Comparisons with Government Reclamation," January 9, 1908, 10 (typescript), Gooding Papers.

33. Louise M. Sill, "The Largest Irrigated Tract in the World," *Harper's Weekly* (October

17, 1908), 52:11 (1st qtn.); Ernest G. Eagleson to Charles Addison Beach, April 5, 1907, Eagleson Papers (2nd qtn.); *Twin Falls News*, May 4, 1906, 3 (3rd qtn.).

34. Samuel P. Hays, *Conservation and the Gospel of Efficiency: The Progressive Conservation Movement, 1890–1920* (Cambridge, MA: Harvard University Press, 1959), 9–10; "A Few Words about the Literature of Irrigation," *National Land and Irrigation Journal* 4 (October 1911), 21; Henry F. Cope, "Making Gardens out of Lava Dust," *World Today* 10 (June 1906):621–28 (1sr qtn.); Montgomery Rollins to State Engineer, December 28, 1909, IRR (2nd qtn.); J. E. Bangs, "Irrigation Securities vs. Other Investments: Pertinent Reasons for High Standing of Irrigation Bonds," *Bonds and Mortgages* (October 1910), copy in IRR (last qtns.).

35. The State Engineer of Idaho, one subordinate of the Land Board, embellished such arguments by claiming that the Carey Act protected investors and settlers from "wildcat schemes" that had historically caused "investment of private capital in irrigation enterprises" to be "fraught with extreme hazard": State Engineer to U.S. Senator Wesley L. Jones of Washington, July 16, 1909, IRR.

36. S. H. Hays to F. A. Voigt, October 13, 1908, Twin Falls Land and Water Company Papers, ISHS (qtns.).

37. Heber Q. Hale to James H. Brady, August 23, 1910, James H. Brady Papers, Governors' Files, ISA; Heber Q. Hale, "Idaho Is 'Uncle Sam's' Greatest Irrigator," *Illustrated Idaho* I (January 1911):12.

38. S. D. Taylor, *Carey Act Projects: Report on the Industrial Development of Idaho, Accompanied by the Reclamation of Desert Lands by Actual Irrigation under the Carey Act* (Boise: n.p., 1913), 5. This official also credited Carey Act reclamation processes with increasing the state's population by 50,000 people during these years. Some of the newcomers invested significant sums of money in their land. In twelve months, one of them reported, he spent $13,000 on developing fruit orchards and might, by selling his holdings elsewhere, spend another $42,000 on his Idaho farmstead: C. E. Holderman to J. H. Hawley, August 1, 1912, Hawley Papers.

39. Williams, *Carey Act in Idaho*, 15; Lovin, "Carey Act in Idaho," 127–128; "All Carey Act Projects," November 1, 1912 (typescript), IRR; F. R. Gooding to Burton L. French, April 16, 1908, French Papers. Histories of a few of the projects have been written. For accounts of projects in the Lost River areas, see Bruce L. Schmalz, "Headgates and Headaches: The Powell Tract," *Idaho Yesterdays* 9 (Winter 1965–1966):22–25; "Footnote to History: 'The Reservoir Would Not Hold Water,'" *Idaho Yesterdays* 24 (Spring 1980):14–19. For accounts of projects at Marysville and Mud Lake, see Scott, "Idaho and the Carey Act," 101–155. Projects developed by Pittsburgh bankers William Kuhn and James Kuhn are treated in Hugh T. Lovin, "A 'New West' Reclamation Tragedy: The Twin Falls-Oakley Project in Idaho, 1908–1931," *Arizona and the West* (Spring 1978), 20:5–24 [Chapter7 in this book]; Lovin, "Free Enterprise and Large-Scale Reclamation on the Twin Falls-North Side Project, 1907–1930," *Idaho Yesterdays* 29 (Spring 1985):2–14 [Chapter 5 in this book]; Lovin, "How Not to Run a Carey Act Project: The Twin Falls-Salmon Falls Creek Tract, 1904–1922," *Idaho Yesterdays* 30 (Fall 1986):9–15, 18–24.

40. Bitton, "Blackfoot," 13. For unsuccessful efforts to begin the long-discussed Dubois project in the same sector of the Snake River Plain, see Scott, "Idaho and the Carey Act," 156–64; State Engineer to S. V. Evers, November 17, 1910, W. G. Swendsen to Herbert A. Thompson, July 30, 1920, and George N. Carter to I. H. Nash, November 12, 1929 and

January 24, 1930, all IRR; J. H. Brady to R. A. Ballinger, May 6, 1909, Brady Papers; W. S. Fulwider to C. Ben Ross, February 3, 1936, File 302.12, BRR.

41. Williams, *Carey Act in Idaho*, 29–30; Taylor, *Carey Act Projects*, 5.

42. Katherine Coman, "Some Unsettled Problems of Irrigation," *American Economic Review* 1 (March 1911):12; Edward Bohm to James H. Hawley, December 30, 1911, Hawley Papers (qtn.); D. N. Niver to James H. Brady, June 9, 1910, and B. B. Brooks to Brady, June 13, 1910, Brady Papers; Granger Farwell to W. M. Wayman, January 26, 1911, and J. B. Sears to Minneapolis Machinery Company, September 26, 1911, King Hill Extension Irrigation Company Papers, ISHS; "Explaining the Pittsburg[h] Crash," *Literary Digest* 47 (July 19, 1913):85.

43. *Carey Act Projects: Report of a Committee Appointed by the Secretary of the Interior to Make an Investigation into and Report upon the History and Present Condition of the Carey Act Projects*, Senate Document 1097, 62d Cong., 3d sess., February 15, 1913 (Washington, DC: Government Printing Office, 1913), 22; Randall R. Howard, "Irrigation Frauds in Ten States," *Technical World Magazine* 17 (July 1912):513; Judson Harmon, "The Tour of the Northwestern Governors: [Remarks] before the Lotus Club, New York, January 6, 1912," in printed press release, January 7, 1912, Hawley Papers (qtns.).

44. Hale, "'Uncle Sam's' Greatest Irrigator," 12; W. G. Swendsen, "Department of Reclamation," December 23, 1920 (typescript), IRR; *Second Biennial Report of the Department of Reclamation, State of Idaho, 1921–1922* (N.p.: n.p., n.d.), 20.

45. As of 1923, the Idaho Commissioner of Reclamation calculated, twenty-three Carey Act projects remained in business, and there were seventy-eight irrigation districts in the state. Within the boundaries of these irrigation districts were 1,723,673 acres: W. G. Swendsen to C. C. Moore, July 20, 1923, Charles C. Moore Papers, Governors' Files, ISA.

46. In discussing the collapse of Idaho's reclamation boom, James Smith (Idaho State Engineer 1915–1919) ascribed little venality to private enterprisers but accused them of many errors like miscalculating the amounts of water that were available at sensible costs. On the other hand, the president of the Idaho Irrigation Congress charged that entrepreneurs behind Carey Act projects included "unscrupulous and cunning promoters" whose wrongdoings persisted because of the state of Idaho's "inadequate laws and want of ordinary horsesense." For his part, a federal engineer judged the Carey Act a law based on the "sound principle" of offering opportunities to free enterprise to profit from its risk-taking, but many Carey Act projects failed because irrigation companies never raised sufficient investment capital: J. H. Smith to F. E. Garrett, July 29, 1916, IRR; *Proceedings of the Joint Conference of Irrigation, Engineering and Agricultural Societies of Idaho* (Twin Falls: Kingsbury Printing Company, n.d.), 92 (1st and 2nd qtns.); D. C. Henny and others, "Some Phases of Irrigation Finance," *Transactions of the American Society of Civil Engineers* (1928), 92:547 (3rd qtn.). Others argued that fundamentally the costs of arid-land reclamation were too high; one canard held that two settlers of arid land must fail before the third one finally succeeded: Roy E. Huffman, *Irrigation Development and Public Water Policy* (New York: Ronald Press Company, 1953), 62.

47. *Third Biennial Report of the Department of Reclamation, State of Idaho. 1923–1924* (N.p.: n.p., n.d.), 17; R. P. Teele, "The Financing of Non-Governmental Irrigation Enterprises," *Journal of Land and Public Utility Economics* 2 (1926):437–38.

48. John M. Haines to F. H. Newell, August 13, 1913, November 28, 1914, and Haines to Franklin K. Lane, January 15, December 17, 1914, John M. Haines Papers, Governors' Files, ISA; *Federal Reclamation by Irrigation: Message from the President of the United States*

Transmitting a Report to the Secretary of the Interior by the Committee of Special Advisors on Reclamation, Senate Document 92, 68th Cong., 1st sess. (Washington, DC: Government Printing Office, 1924), 166; Olive Groefsema, *Elmore County: Its Historical Gleanings* (Caldwell, ID: Caxton Printers, 1949), 371; Minutes of State Board of Land Commissioners, December 17, 1917, 1–3, IRR.

49. "Minidoka Project History, 1917," 68 (typescript), Project Histories, Minidoka, Roll 99; F. Ross Peterson and W. Darrell Gertsch, "The Creation of Idaho's Life Blood: The Politics of Irrigation," *Rendezvous* 11 (Fall 1976): 58; Hugh T. Lovin, "'Duty of Water' in Idaho: A 'New West' Irrigation Controversy, 1890–1920," *Arizona and the West* (Spring 1981), 23:13–25 [Chapter 3 in this book]; "Status of Carey Act Projects," *Sunset Magazine* (1919), as cited in W. G. Swendsen to *Sunset Magazine*, November 15, 1919, IRR (1st qtn.); John Haines to E. E. Elliott, November 25, 1914, John Haines Papers, Personal Correspondence Files (2nd qtn.).

50. Fred R. Reed, "Watch Your Step," n.d. (mimeographed), IRR (qtns.); John F. Nugent to Ernest G. Eagleson, August 13, 1919, Eagleson Papers; Gertsch, "Upper Snake River Project," 177–187. The Idaho Commissioner of Reclamation argued that the "easy projects" were under way and that about 2,250,000 acres awaited "future development"; also, he speculated, with less water loss from canals, more efficient irrigation systems, and "more care in application of water," altogether about 6,500,000 acres could be irrigated: W. G. Swendsen to C. C. Moore, October 8, 1925, Moore Papers. The Reclamation Service was reorganized and renamed the Bureau of Reclamation in 1923. The bureau's first head, or commissioner, was former Idaho governor D. W. Davis.

51. Harold Conkling to Chief of Construction, January 26, 1920, Engineering Correspondence File 201-i, BRR; G. N. Carter to M. H. Barton, October 16, 1923, IRR; Williams, *Carey Act in Idaho*, 68–69, 74.

52. W. G. Swendsen to Rhea Luper, July 18, 1924, IRR (1st qtn.); "Eighth Annual Meeting of Western State Engineers," November 13–14, 1935, 13–14 (typescript), C. Ben Ross Papers, Governors' Files, ISA (2nd qtn.). About a decade later, Governor Charles Robins said that, with cheap and easily constructed irrigation projects nearing completion, the "trend" in the state was away from "explotion [*sic*] of development to conservation of our resources": C. A. Robins to A. C. Pierson, n.d., IRR.

53. John M. Boyle to W. G. Swendsen, March 22, 1920; clipping from *Arco Advertiser*, August 10, 1925; J. M. Lampert, "History of Salmon River Diversion: Supplemental Water Supply for the Boise Valley—The Necessity for It and Its Feasibility—Losses Resulting from Inadequate Supply," n.d. (typescript); and R. W. Faris, "Supplementary Water for Irrigation in Idaho, with Particular Reference to [the] Boise and Snake River Valleys," n.d. (typescript), all in IRR; C. Ben Ross to Chase Clark, September 6, 1935, and Ross to R. F. Walter, September 25, 1935, Ross Papers; John Ise, *Our National Park Policy: A Critical History* (Baltimore: Resources for the Future, Inc., Johns Hopkins University Press, 1961), 307–17, 433–34; Merrill D. Beal, *The Story of Man in Yellowstone* (Caldwell, ID: Caxton Printers, 1949), 259; Aubrey L. Haines, *The Yellowstone Story: A History of Our First National Park*, 2 vols. (Yellowstone National Park, WY: Yellowstone Library and Museum/Colorado Associated Universities Press, 1977), 2:333–334, 342–345.

54. Gertsch, "Upper Snake River Project," 169*ff*; Irvin E. Rockwell, *The Saga of American Falls Dam* (New York: Hobson Book Press, 1947); *Sixth Biennial Report of the Department of Reclamation, State of Idaho, 1929–1930* (N.p.: n.p., n.d.), 15–20; *Idaho Statesman* (Boise), March 25, 1990, 1A, 8A, and July 19, 1990, 4C; *Project Data 1981*, 641.

55. Peterson and Gertsch, "Creation of Idaho's Life Blood," 60; *Project Data 1981*, 641–642, 643, 747; Lynn Crandall to B. E. Stoutemeyer, July 9, 1936, Engineering Correspondence File 738, BRR.

56. *Project Data 1981*, 43–47; Beal and Wells, *History of Idaho*, 2:187; Susan M. Stacy, "Economic Development and Federal Flood Control on the Boise River, 1943–1985" (master's thesis, Boise State University, 1989); Carlton, "Boise Irrigation Project," 94–96; *Fourth Biennial Report of the Development of Reclamation, Stare of Idaho, 1925–1926* (N.p.: n.p., n.d.), 30–31; J. H. Lowell, "Report to the Directors of [the] Gem Irrigation District" (printed booklet, February 7, 1928), James H. Lowell Papers, ISHS.

57. Carlos A. Schwantes, *The Pacific Northwest: An Interpretive History* (Lincoln: University of Nebraska Press, 1989), 204.

58. *Idaho Almanac, 1977 Edition* (Boise: Executive Office of the Governor and the Idaho Division of Tourism and Industrial Development, [1977]), 171–72, 199; Opie, *Law of the Land*, 111–18; Richard Lowitt, *The New Deal and the West* (Bloomington: Indiana University Press, 1984), 90, 94, 146. Consideration of the federal dam on the Teton River began in 1933, but the plan was shelved at that time to pacify water-users who insisted that the next federal dam be constructed on the South Fork of the Snake River: Elwood Mead to R. W. Faris, October 14, 1933, and Lynn Crandall to James Pope, October 20, 1933, IRR.

59. *Idaho Almanac, 1977 Edition*, 172; Leonard J. Arrington, "Irrigation in the Snake River Valley: An Historical Overview," *Idaho Yesterdays* 30 (Spring/Summer 1986):8–9; *Project Data 1981*, 642; *Idaho Statesman*, September 26, 1990, 8B.

60. *Idaho Almanac, 1977 Edition*, 177–178.

61. G. W. Lineweaver to M. R. Kulp, March 14, 1946, IRR (1st qtn.); *Project Data 1981*, 641, 643; "Old Reliables Are Feelin' Their Age," *Idaho Statesman*, February 19, 1984, 2F, 3F; William F. Ringert, "Irrigation Districts: Purpose, History, Funding, and Problems," *Idaho Yesterdays* 30 (Spring/Summer 1986):66, 68; *Idaho Statesman*, August 29, 1990, 1C, 2C (all other qtns. on 2C).

62. Lowitt, *New Deal and the West*, 224–228; Merle Wells and Arthur A. Hart, *Idaho: Gem of the Mountains* (Northridge, CA: Windsor Publications, 1985), 137–139, 141, 209, 220; Peterson, *Idaho*, Ch. 9; Scott W. Reed, "The Other Uses of Water," *Idaho Yesterdays* 30 (Spring/Summer 1986):33–44. For an especially helpful source on regional western development, see Gerald D. Nash, *World War II and the West: Reshaping the Economy* (Lincoln: University of Nebraska Press, 1990), especially Ch. 8.

63. Opie, *Law of the Land*, 125–26; Richard Lee Reid, "The Analysis of Irrigated Agricultural Development, Mountain Home Division, Southwest Idaho Water Development Project" (M.S. thesis, University of Idaho, 1973), 3–6; Robert G. Dunbar, *Forging New Rights in Western Waters* (Lincoln: University of Nebraska Press, 1983), 150–51; Randy Stapilus, *Paradox Politics: People and Power in Idaho* (Boise: Ridenbaugh Press, 1988), 297–303; *Idaho Statesman*, July 24, 1990, 1A, 6A, July 26, 1990, 1A, 10A, August 25, 1990, 1C, 3C, October 12, 1990, 1C, November 10, 1990, 1C, 2C.

Remembering My Father

Hugh and Ida Lovin with son, Jeffrey, 1962

My father, Hugh Taylor Lovin, grew up on a ranch in southeast Idaho near Inkom. Lessons in working hard and saving, taught for survival in the Great Depression, were ingrained during a childhood spent on a corner of his grandfather's ranch, hewn out of original homesteader claims. In 1935 his family moved to a ranch on Marsh Creek. Like so many Idaho streams, Marsh Creek's water was not abundant and had to be carefully dammed and conserved for the survival of adjacent crops and livestock. Snowfall was fickle, and unless dammed the precious snowmelt would rush away to the Snake, the Columbia, and finally the Pacific Ocean. Dad dug ditches and small diversion dams to shepherd water onto fields of wheat, alfalfa, and oats. It was hot, dirty work. Yet when the August sun shone, the fruits of his labors showed in bronzed fields swaying with the wind, every bushel a precious contribution to his family's survival.

During long, harsh winters dim kerosene lamps allowed my father to read in a tiny home without electricity or indoor plumbing. When nature called, he made a dash to the outhouse, even during winter blizzards. Running water was found in the creek, tens of yards from their small, rough-wood home. Indeed, their refrigerator was a box with a hole through which cold water from the creek ran and which kept milk and foods from spoiling. Dad's chores started at 5 a.m. on a stool beside the cows. He went to a local school eight miles away, came home to help brand cattle, herd sheep, and keep the coyotes at bay. A shotgun stood next to their kitchen door for that purpose, as well as a shovel for rattlesnakes that lurked amongst the lava rocks. A family of five filled the small house, yet Dad would stay up late gathering knowledge from books lent by the small Andrew Carnegie library in town. Roosevelt's New Deal failed to give relief to their hardscrabble life, but Dad's own industry helped reap the knowledge and wisdom books can provide. Despite a youth in the fields and on horseback, Dad always placed high value on the need to read, learn, and think.

The land, the water, and his experiences in managing both helped guide Dad's observations of water scarcity in the arid West and its effect on local politics. Neighbors argued, and indeed fought over water in order to survive. People had to work together to find solutions with what little water was there, using primitive equipment they could afford. Opinions varied widely, and voices grew louder during droughts. Half-baked opinions influenced elections. Dad observed the often nasty exchanges and, as he told me, thought public discourse needed to be raised so that intellect rather than shouting might solve problems. And so he left the ranch, saving money for college by mixing lime and clinker to make cement at the Inkom cement plant, loading trucks with fruit, and driving a bus while he worked his way through Idaho State University (at that time Idaho State College) in Pocatello. He lived with his grandmother and earned money teaching school in southern Idaho. Lessons learned in work and in reading, of pulling up your own bootstraps and of compassion for the struggles of other less fortunate, led him to believe in the need for decency in caring for our fellow man. From Idaho history he recognized the plight of Idaho silver miners seeking decent wages, the right of unions to seek fair treatment, the Wobblies and Socialists—how they all jockeyed for power and offered sometimes ridiculous solutions to win votes and political power. Could not these lessons of history provide a more logical scaffolding upon which to find solutions?

Dad earned a master's degree in history at Washington State University, and PhD at the University of Washington. He spent his teaching career at what is now Boise State University. A third generation Idahoan, he desired to give back to his home state. He researched and published constantly while teaching.

Though no chemist, Dad told me once of his amazement that such a small molecule, two measly hydrogen atoms bonded to one oxygen atom, formed the fundamental commodity on earth. Dad warned me that the conflicts he had studied in the river valleys of Idaho would someday be replicated on a larger basis—nationally and globally. Individuals and political entities will fight for that tiny H2O molecule. Dad wrestled to blend disparate facts culled from state archives, interviews with stakeholders, political records and statehouses, and field research, to amalgamate an accurate, unbiased view of how water issues developed. He then transferred that understanding one keystroke at a time into his manual typewriter in our small living room. Dad hoped his first-person research, much of it unpublished, and articles would help historians and legislators avoid repetition of past fruitless conflicts over water.

Dad served as chairman of the BSU History department. He enjoyed his students. But research, writing, and presenting papers never stopped. I may have wished for more time throwing the football with him, but history was his passion. Scholarship was his intent. Dad was stern, and had high standards for his students, for society, and for me. Yet he was also a generous father. He provided me with a trumpet and piano. He bought items for other activities, like my martial arts robe and a basketball to play the game I loved. But he emphasized success and placed education as the highest priority, and shuttled me off to the Ivy League.

Dad also had a softer side. Returning home from Cornell during college break, I learned my Dad had taken sympathy on a college student from a small Idaho town who, though he sat in the front row, could never seem to get above a C, and sometimes sadly D's, on his tests. Dad knew the student was trying, indeed trying very hard given in part his rural background like Dad's. But what Dad saw was industry and self-motivation, and no request for a handout. The student quietly worked, wrote, and rewrote his notes. To my utter shock from an otherwise disciplined father, Dad gave the student a final grade of B. I shall never forget his parting comment, "Son, never forget the power of honest work, the inherent ability of your own self to achieve, and beauty of kindness to others." To Dad, the key to success for each of us is to study and learn. May this collection of essays serve as an example of his efforts to help us all.

Jeffrey D. Lovin
AB Cornell, MD University of Washington

Postscript: Mom, a schoolteacher, supported and deeply loved Dad. A gracious woman, she too loved knowledge. Mom and Dad always had a pile of books from the local library waiting for my attention after regular schoolwork. Dad added more to the pile as it shrank. He told me "acquisition of knowledge is the beginning of wisdom, and wisdom is the beginning of civility." His grandchildren, whom he deeply loved, have honored his love of learning: Kristen Elizabeth Lovin, AB Harvard, JD Columbia; Erika Marie Lovin, AB Harvard, MBA Wharton, MPA Harvard; Benjamin Daniel Lovin, AB Cornell, MD Wake Forest.

Hugh Lovin and family, 2014. Front, Benjamin Daniel, and behind from left Erika Marie, Hugh, Kristen Elizabeth, and Jeff Lovin. *Photo by Kristen Elizabeth Lovin.*

Index